# Concluding Papers of Yuri Mnyukh on Solid-State Physics

Mechanism and Kinetics of Phase Transitions and Other Reactions in Solids

Second-Order Phase Transitions, L. Landau and his Successors

On Phase Transitions that Cannot Materialize

Hysteresis of Solid-State Reactions: its Cause and Manifestations

Phase Transitions in Layered Crystals

On Physics of Magnetization

Ferromagnetic State and Phase Transitions

Magnetization of Ferromagnets

Paramagnetic State and Phase Transitions

The True Cause of Mgnetostriction

The Physical Nature of "Giant" Magnetocaloric and Electrocaloric Effects

The Nature of Ferroelectricity

Superconducting State and Phase Transitions

Searching for a Critical Phenomenon

**CONCLUDING PAPERS OF YURI MNYUKH ON SOLID-STATE PHYSICS**

First Printing: 2017 as "Concluding Papers of Yuri Mnyukh on Phase Transitions"

Second Printing: 2020

ISBN 978-0-578-70177-6

DirectScientific Press
Farmington, CT 06032

www.mnyukh.com

# Introduction

The articles of this book, dated between 2011 and 2020, represent the final results of an investigation of solid-state phase transitions undertaken by Yuri Mnyukh in 1959 after receiving a doctorate degree for his work on the phase behavior of paraffins (published in J. Phys.Chem. Solids 24, 631-640, 1963). The purpose of the new study was to reveal how a crystal structure changes into a different structure with temperature. That was unknown in those times, hence approach to the subject depended on an opinion. The most popular opinion was based on the belief that it ("of course") is a cooperative molecular displacements/deformations/distortions. The other, dominated among theorists, divided the phase transitions into "first-order"and "second-order" but regarded both to be a sudden cooperative switch at a "critical" point. The third claimed that phase transitions proceed through an intermediate amorphous or vacuum layer between the phases. Neither approach was able to account for the phenomenon adequately.

A number of key experimental discoveries (described in the first article of the book) by Mnyukh and his graduate students have shown that the process is always a crystal growth. It involves a peculiar contact structure of the interface and the nucleation mechanism not predicted by any theory. From these experiments the nucleation-and-growth molecular mechanism of phase transitions was deduced. Replacement of the initial structure proceeds not cooperatively, but by molecule-by-molecule relocation at the interfaces.

The results of the Mnyukh's investigations far exceeded their original purpose. The nucleation-and growth mechanism was extended to cover other solid-state reactions. It turned out to be the universal tool to unravel the physical nature of several basic phenomena of solid-state physics. The second-order phase transitions were shown to be nonexistent and the detrimental effects of their theory by Landau on solid-state physics were made evident. The origin of ferromagnetism was accounted for. The molecular mechanism of ferromagnetic transitions, as well as of a magnetization process was explained. The fallacy of the Heisenberg theory of ferromagnetism was demonstrated. The origin of ferroelectricity and its close analogy to ferromagnetism was uncovered. The mechanism of phase transitions in superconductors was specified. The great enigma of theoretical physics – "heat capacity lambda-anomalies" in solids and liquid He – was eliminated. Finally, the overall conclusion has been made regarding viability of the whole theory of *critical phenomena*, considering that it lacks a real-world application. The reader will find all that and more in this book.

Since the articles have been written to be each self-consistent, certain repetitions in this book are inevitable, caused by the need to explain the nucleation-and growth mechanism, lambda-peaks, or thermodynamics again. Altogether, the collected articles are not a "mainstream" in the current literature. Inertia is strongly embedded in scientific theoretical circles, cemented in the present case by the high authority of Landau, the author of the theory of second-order phase transitions, and Heisenberg, the author of the theory of ferromagnetism. The reader of this book will find that these theories do not approximate reality. Nevertheless, they are taught in the universities worldwide to this day. An untold truth is that a theory by renowned author, if resulted from a pure creativity without care to match the available evidence, can stifle, rather than promote scientific advancement. And this is the case with the theories in question. It will take time, but the solutions presented in this book are well substantiated, logically monolithic, answering all questions left by conventional theories. They are destined to prevail.

V. J. Vodyanoy
Professor, Director of Biosensors Laboratory
Auburn University

# CONTENTS

# Mechanism and Kinetics of Phase Transitions and Other Reactions in Solids

The work leading to the universal *contact* molecular mechanism of phase transitions and other reactions in solids is presented. The two components of the mechanism - nucleation and interface propagation - are investigated in detail and their role in kinetics is elucidated. They were shown to be peculiar: nucleation is "pre-coded", rather than resulted from a successful fluctuation, and the interface propagates not by frontal advancement as a whole, but by molecular filling thin layers in the transverse direction. The type of the nucleation sites is determined. An inherent instability and irreproducibility of the kinetics is revealed. A *linear kinetics*, as opposed to the bulk kinetics, is shown to be in accord with the *contact* mechanism. Ferromagnetic phase transition and magnetization process are added to the list of solid-state reactions: neither occurs without nucleation-and-growth structural rearrangement.

American Journal of Condensed Matter Physics 2013, 3(4): 89-103
DOI: 10.5923/j.ajcmp.20130304.01

# Mechanism and Kinetics of Phase Transitions and Other Reactions in Solids

Yuri Mnyukh

Chemistry Department and Radiation and Solid State Laboratory, New York University, New York, NY 10003, USA

**Abstract**   The work leading to the universal *contact* molecular mechanism of phase transitions and other reactions in solids is presented. The two components of the mechanism - nucleation and interface propagation - are investigated in detail and their role in kinetics is elucidated. They were shown to be peculiar: nucleation is "pre-coded", rather than resulted from a successful fluctuation, and the interface propagates not by frontal advancement as a whole, but by molecular filling thin layers in the transverse direction. The type of the nucleation sites is determined. An inherent instability and irreproducibility of the kinetics is revealed. A *linear kinetics*, as opposed to the bulk kinetics, is shown to be in accord with the *contact* mechanism. Ferromagnetic phase transition and magnetization process are added to the list of solid-state reactions: neither occurs without nucleation-and-growth structural rearrangement.

**Keywords**   Phase Transitions, Solid-State Reactions, Nucleation, Interface, Kinetics, Crystal Growth, Crystal Defects, Molecular Mechanism

## 1. Introduction

The basics of solid-state phase transitions[1-16] will be summarized and supplemented here by analysis of their kinetics − an aspect of essential theoretical and applied importance. The term *kinetics* means relationships between the macroscopic rate of a phase transition and any conditions or parameters it depends on. The notion *kinetics* implies *phase coexistence*, for it makes sense only if the mass fraction between the two phases is changing during the transition. The only conceivable way of this change is *nucleation and propagation of interfaces*. In other words, investigation of kinetics of phase transitions already means recognition of their nucleation-and-growth mechanism. Only two ways to materialize for crystal phase transitions are conceivable: (a) by nucleation-and-growth and (b) by instant change at a critical point[1,2]. If a phase transition is a "critical phenomenon", its kinetics must not exist, for it comprises all the matter at once as soon as its critical temperature $T_c$ is attained. The rate of this instant transition is function neither time *t*, nor temperature T. The concepts "kinetics" and "critical phenomenon" are incompatible. It should be noted that instant phase transitions have not been found[1,2].

Kinetics of phase transitions is inseparable from their molecular mechanism. As soon as the nucleation-and-growth

nature of solid-state phase transitions is recognized, it becomes evident that comprehension of their kinetics requires a certain knowledge of the structure and properties of the nuclei and interfaces. Critical step in that direction has been a discovery of *edgewise* (or *stepwise*) molecular mechanism of phase transitions[5,7,10]. The nucleation, in particular, was found quite different from the "classical" interpretation, its features critically affecting the phase transition kinetics, as will be demonstrated.

Closely preceding to that discovery was the 1960 *International Symposium on Reactivity of Solids*[17]. It "focused[its] attention on the mechanism and kinetics of reaction in solids". It dealt with such reactions as polymorphic phase transitions, recrystallization, decomposition, chemical reduction and polymerization, not counting those involving also liquid or gas. The vast literature that treated almost all phase transitions as "continuous" and "critical phenomenon" was noticeably ignored. It was observed that "imperfections are preferred sites for internal nucleus formations". (The term "preferred" means, however, that possibility of a *homogeneous* nucleation does not eliminated). These imperfections are: vacancies, interstitials, foreign atoms or ions, linear dislocations, screw dislocations. Their interaction and diffusion were discussed. It was noted that the more perfect crystal, the lesser is its reactivity. The term "nucleation and growth" was common. It was not new, however, for in 1930's and 1940's it was used in developing of what can be called "bulk kinetics" theories. Nevertheless, nucleation-and-growth phase transitions were not considered the only way to occur. For example, one contributor[18] classified solid-state

* Corresponding author:
yuri@mnyukh.com (Yuri Mnyukh)
Published online at http://journal.sapub.org/ajcmp

transformations as *topotactic, epitactic* and *reconstructive*, only the last one to occur by nucleation and growth. In another case[19] the imaginary way of the olivine-spinel restructuring through intermediate states, assisted by dislocations and diffusion, was proposed. It was not mentioned how the notion "kinetics" can be applied to the phase transitions assumed not to occur by nucleation and growth.

In the *bulk kinetics* the mass fraction *m* of one phase in a two-phase specimen was the value of interest. Its rate depends both on multiple nucleation and growth in unknown proportions, different in every particular case. Nucleation critically depends on the presence, distribution and generation of specific lattice defects. It is quite different in a perfect and imperfect single crystal, in a big and small crystal, in a single crystal and polycrystal, in a fine-grain and coarse-grain polycrystal or powder. Growth is not a stable value either, not repeating itself, for example, in cycling phase transitions. The nucleation and growth, when they act together, are not only irreproducible and uncontrollable, there is no way to theoretically separate their contributions in order to calculate the total bulk rate. In this context, the theoretical approach called "formal kinetics" should be mentioned where an attempt to separate them was undertaken. The Avrami work[20] is most known. One of his main assumptions − isothermal rate of nucleation − was invalid due to a "pre-coded" character of both nucleation sites and the nucleation temperatures[16]. Bulk kinetics can shed no light upon the physics of phase transitions or properly account for their kinetics. As a minimum condition, the nucleation and interface motion contributions must be experimentally separated. The best way to do that is a visual observation of nuclei formation and interface motion in optically transparent single crystals. This is a method of *interface kinetics*. The purpose is to reveal its physics, rather then phenomenology.

As mentioned, there is a reproducibility problem. The absolute velocity V of interface motion is not reproduced even in the same single crystal. Fortunately, valuable information can be obtained from a *relative* V changes. As will be shown, it helps not only verify and substantiate the suggested *contact* mechanism, but to penetrate deeper into its details. It will be demonstrated that the universal *contact* mechanism accounts for all the complexity, versatility, and poor reproducibility of the kinetics of solid-solid phase transitions.

## 2. Interface and Its Motion

After the discovery has been made that phase transitions in single crystals of *p*-dichlorobenzene (PDB) and some other substances were a growth of well-bounded single crystals of the new phase (Fig. 1)[4,6] and that orientation relationship (OR) between the initial and resultant crystals did not exist[6], the attention was concentrated on the mode of interface propagation. Observation of the interface was undertaken under maximum resolution attainable with a regular optical microscope. It was found that its motion has *edgewise* (or "*stepwise*") mechanism[5,7]. Its advancement in the normal direction proceeded by transverse shuttle-like strokes of small steps (kinks), every time adding a thin layer to it (Fig. 2). That was the same mechanism of crystal growth from liquids and gases[21,22], going down to the molecular level. A generalization came to light: any process resulting in a crystal state, whichever the initial phase is − gas, liquid, or solid − is a *crystal growth*. It proceeds by the edgewise molecule-by-molecule formation of layers and layer-by-layer additions to the natural crystal face. While crystal growth from gaseous and liquid phases is called *crystal growth*, and crystal growth from a solid phase is called *phase transformation* or *phase transition*, the difference is semantic. This does not mean, of course, that crystal growth in a crystal medium does not have its specificity.

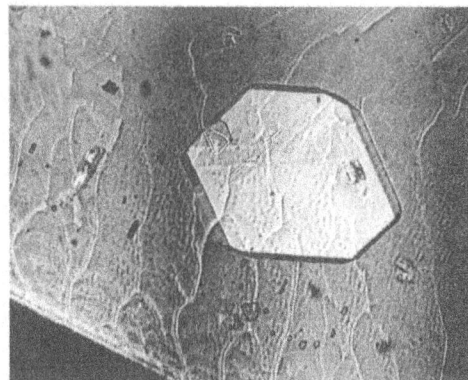

**Figure 1.**    Growth of well-bounded H single crystal in the single-crystalline PDB plate. A part of the natural edge of the L crystal is visible at the left lower corner. The real diameter of the H crystal is 0.4 mm. (L and H are low- and high-temperature phases respectively)

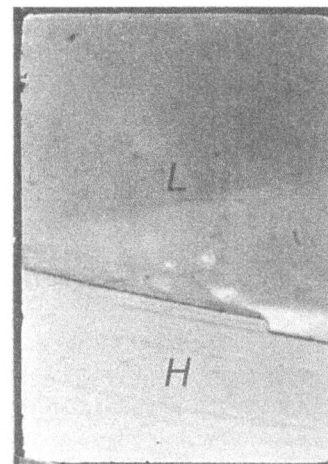

**Figure 2.**   L → H phase transition in PDB. The 2.5 μm high "kink" (step) is moving from left to right. The kink is, in fact, an avalanche of molecular steps

The physical model of a solid-solid interface and the manner of molecular rearrangement at this interface was needed. A model assuming existence of a vapor gap between the phases was tried and ended in impasse by Hartshorne and colleagues[23-35]. They concluded that the activation energies of solid-state phase transitions $E_A$ and the heat of sublimation $E_S$ were equal, but the velocity of interface movement was $10^3$ to $10^5$ times higher than can be provided by evaporation into the gap ("Hartshorne's paradox"). On the other hand, any amorphous interlayer of excited molecules cannot exist either: phase transition is localized only at the kinks, and there would be plenty of time for the excited molecules, if any, of the smooth interface to join the stable phase.

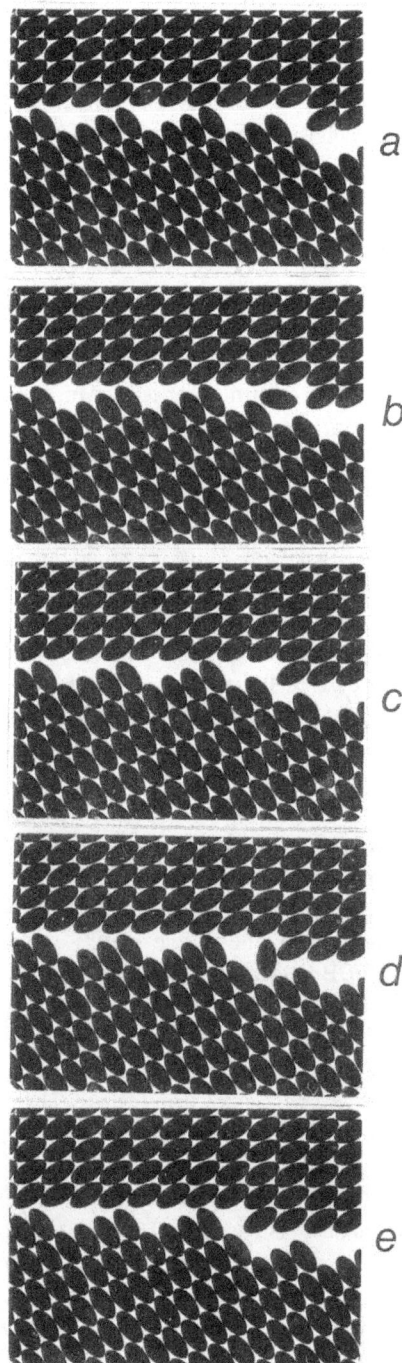

**Figure 3.** Molecular rearrangement at the *contact* interface during phase transition (frames of the animated film). The effective gap between the phases is 0.5 molecular layer. Building up (a → e) of a new molecular layer

The *contact* interface, shown in Fig.3 with a 2-D model, meets all the observations. There is neither vacuum, nor amorphous transitional layer. Two crystal phases are simply in contact with each other, coupled by the molecular forces. The *contact* structure of the interface requires neither any correlation between the structural parameters of the two phases, nor a particular OR. Interface on the side of the resultant crystal consists of molecules closely packed into a layer of low crystallographic indices $(h, k, \ell)$. In relation to the initial lattice this direction can be irrational. There is a net of microcavities at the interface that cannot accommodate additional molecules. No essential lattice distortions exist.

The *contact* interface was subjected to a number of additional tests[36,7]:

♦ *Coupling at the contact interface (observation)*. While the Hartshorne's teem worked with polycrystal films, our dealing with single crystals allowed immediately reject the "vapor gap" model. Indeed, handling a "hybrid" crystal crossed by a flat interface did not result in its separation to two detached parts of individual phases, as it would be in the case of a vapor gap between the phases. Consequently, there is an essential coupling at the interface. On the other hand, the interface is the weakest section, as expected from the its *contact* model. Application of an external force breaks the "hybrid" crystal exactly along the interface in two separate parts.

♦ *Coupling at the contact interface (measurement and calculation)*. The coupling force between the phases at their interface was measured by breaking the "hybrid" crystal by a microdynamometer and then compared with the calculated value for the phases separated by 0.5 molecular layer as suggested by the model. The two values turned out to be in a reasonable agreement.

♦ *Solution of the Hartshorne's paradox*. Molecular rearrangement at the *contact* interface is a one-step molecular relocation under the attractive action of resultant phase. This process can be termed "stimulated sublimation". Accordingly, $E_A$ is lower than $E_S$ by the attraction energy $E_{attr}$. For the effective gap 0.5 molecular layer, the computer calculations led to

$$E_A = \sim 0.7\ E_S.$$

When believing that $E_A = E_S$, the Hartshorne's team was not far from truth, but the above 30% difference accounts for the observed $\sim10^4$ times faster phase transition as compared to sublimation.

Fig. 3 also shows how the interface advances by molecule-by-molecule process. The frame sequence illustrates an *edgewise* movement of a molecular step. Molecules detach from one side to build up a closely packed layer on the opposite side. Once one layer is completed, the interface becomes advanced by one interlayer spacing, while the "contact" structure of the interface at its new position is preserved. For the phase transition to continue, a new nucleus has to form on the interface. Here the specificity comes about. Formation of a 2-D nucleus on the flat regions of the contact interface requires extra free space, such as a vacancy or, possibly, their cluster. If it is available, a new molecular step can form to run along the interface. This vacant space will accompany the running step, providing for steric freedom just where the molecular relocation is to occur at the moment. The vacant space will ultimately come out on the crystal surface. Formation of a new step will require another vacant space residing at the new interface position or migrating to it. The *presence of these crystal defects in sufficient quantity is a necessary condition for a phase transition to proceed.* Their availability and diffusion are major controlling factors of the phase transition kinetics

# 3. Nucleation in Crystals

Nucleation is one of the two elements - nucleation and growth - constituting the phenomenon of solid-state phase transitions. Nevertheless, there is vast theoretical literature that treats them as if the phenomenon of nucleation (and subsequent growth) is nonexistent. For example, nucleation is missing in the three (unrelated) books *Structural Phase Transitions*[37-39], in many other books on the subject[*e. g.*, 40-47], in most volumes of *Phase Transitions and Critical Phenomena*[48], and in innumerous journal articles. Yet, some literature on nucleation in a solid state exists, mainly owing to needs of solid-state reactivity, as described above, and physical metallurgy[49,50]. In the latter case the theory of nucleation in a liquid phase was slightly modified to cover solid state[49,51]. But no one previously experimentally verified its theoretical assumptions by using the simplest and most informative objects – good quality small transparent single crystals.

The following data were accumulated with thermal-induced phase transitions in tiny (1 - 2 mm) good quality transparent organic single crystals under controlled temperatures, PDB being the main object[16]. *Notations*: L and H are low- and high-temperature phases; $T_o$– temperature when free energies of the phases are equal, $T_m$– temperature of melting, $T_n$– temperature of nucleus formation. In PDB, $T_o = 30.8°C$ and $T_m = 53.2°C$.

*Nucleation requires finite overheating / overcooling.* Nucleation never occurs at $T_o$ or closer than at a certain finite distance from it. The "prohibited" range for PDB is at least 28 to 32°C. Upon slow heating, nucleation in most PDB crystals will not occur until the temperature exceeds 38°C. Often L-phase melts at $T_m$ without transition into H.

*Nucleation is a rare event.* Slow heating (*e. g.*, 1 to 10°C per hour) does not produce many nuclei. Usually there are only a few units, or only one, or no nucleation sites at all. This observation is at variance with the notion "rate of nucleation" and the statistical approach to nucleation - at least as applied to a single-crystal medium.

*Exact temperature of nucleation $T_n$ is unknown a priori.* Formation of a nucleus upon slow heating of a PDB single crystal will occur somewhere between 34 and ($T_m$=) 53.2°C, and in some cases it would not occur at all. The $T_n$ vary in different crystals of the same substance. .

*Only crystal defects serve as the nucleation sites.* In other words, the nucleation is always *heterogeneous* as opposed to the *homogeneous* nucleation assumed to occur by a successful fluctuation in any point of ideal crystal lattice. The location of a nucleus can be foretold with a good probability when the crystal has a visible defect. Nucleation in sufficiently overheated / overcooled crystals can be initiated by an "artificial defect" (a slight prick with a glass string). In such a case, a nucleus appears at the damaged spot. In cyclic phase transitions L $\rightarrow$ H $\rightarrow$ L $\rightarrow$ H the nucleus frequently appears several times at the same location.

T*he higher crystal perfection, the greater is overheating or overcooling* $\pm \Delta T_n = T_n - T_o$. In brief, better crystals exhibit wider *hysteresis,* for $\pm \Delta T_n$ *is* a hysteresis[13]. The correlation between the degree of crystal perfection and $\Delta T_n$ is confirmed in several ways. One, illustrated with the qualitative plot in Fig. 4, is the $\Delta T_n$ dependence on the estimated quality of the single crystals. Different grades were assigned to sets of crystals depending on their quality, higher grades corresponding to higher crystal perfection. The grades were given on the basis of crystal appearance (perfection of the faces and bulk uniformity) and the way the crystals were grown (from solution, from vapor, rate of growth, temperature stability upon growing, etc.). For the solution-grown PDB crystals the graduation exhibited good correlation with the levels of $\Delta T_n$ within the grades 1 to 4. Two findings will be noted regarding $\Delta T_n$ in these crystals:

1. $\Delta T_n <1.9°C$ was not found even in the worst grade (grade 1) crystals, indicating existence of a minimum (threshold) overheating required for nucleation to occur.

2. The most perfect among the solution-grown crystals were incapable of L$\rightarrow$H changing, so it was L that melted. There was no doubt in the availability of dislocation lines, individual vacancies, interstitial and foreign molecules in those crystals, but they were not of the "proper" nucleation type. Yet, an "artificial defect", created by a slight prick at the temperature near $T_m$, immediately initiated the phase transition from the damaged spot.

$T_n$ *is pre-coded in the crystal defect acting as the nucleation site.* Fig. 5 shows the results of microscopic observation of PDB (L) single crystals upon slow heating. $T_n$ was recorded as soon as nucleation of the H phase was noticed. Growth of the H crystal was quickly stopped and the specimen was returned to the L phase. The cycle was repeated with the same crystal many times. Then the whole procedure was performed with another crystal. The experiments revealed that (a) a nucleus appears every time at exactly the same location, (b) the $T_n$ repeats itself as well, and (c) a particular $T_n$ is associated only with the particular nucleation site. A different $T_n$ is found in another crystal, also associated with the defect acting as the nucleation L site. Thus, *every crystal defect acting as a dormant nucleation site contains its own $T_n$ encoded in its structure.* If a crystal has more than one dormant nucleation site and is subjected to very slow heating, only one site with the lowest $T_n$ will be activated. Upon faster heating, the second nucleus of the second lowest $T_n$ may have time to be activated before growth from the first nucleus spreads over its location…and so on.

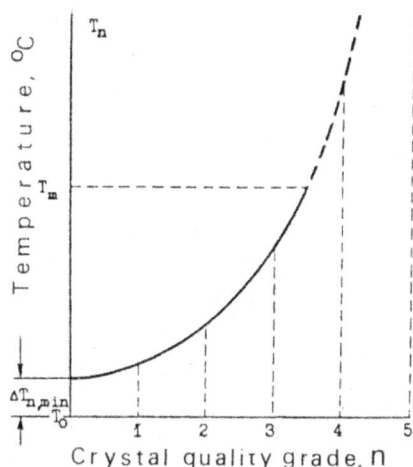

**Figure 4.** The character of dependence of nucleation temperature $T_n$ on quality of single crystals; (L→H). The quality is represented by number $n$: the higher estimated crystal perfection, the higher number $n$. No formation of a nucleus is possible until overheating exceeds some threshold value $\Delta T_{n,min}$. Overheating $\Delta T_n = T_n - T_0$ increases with $n$. Nucleation in "grade 4" crystals does not occur spontaneously, no matter how long they are stored just below melting point $T_m$. Yet, "grade 4" can be coerced into phase transition by a slight prick. No way has been found to induce a phase transition in "grade 5" crystals

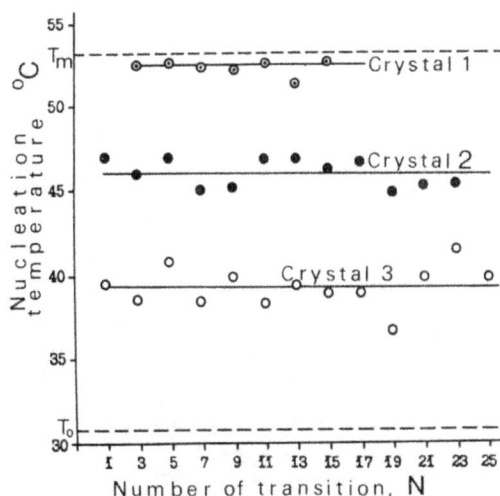

**Figure 5.** Reiterative formation of a nucleus in small PDB single crystals. Reiteration of nucleation temperature $T_n$ when the nucleus repeatedly forms at the same site in the cyclic process. The procedure was as follows. Temperature of a microscope's hot stage with PDB single crystal was slowly raised. As soon as a nucleus of H phase became visible, $T_n$ was measured and the phase transition was immediately reversed. In the subsequent cycles the nucleus formed at the same site (as a rule, at a visible defect) and at the same $T_n$. The measurements for three single crystals are shown. Every particular site was associated with its particular $T_n$. N is ordinal number of phase transition

*Orientation of the resultant crystal is also pre-coded in the crystal defect acting as the nucleation site.* The crystals of new phase in PDB phase transitions grow in random orientations. As long as a nucleus forms at the same lattice defect, the 3-D orientation and even the shape of the growing

crystal repeats itself. Another nucleation site would produce a different orientation inherent exclusively to it. Thus, information on the orientation of the new crystal is pre-coded and stored in the structure of the crystal defect serving as the nucleation site. This information, however, remains dormant unless this defect is activated (by heating or cooling to the $T_n$ encoded in the defect).

*The $T_n$ in the same crystal can be changed.* For example, in case of L→H phase transitions, the $T_n$, initially encoded in a nucleation site, can be "erased" by creating an "artificial defect", as mentioned earlier. The $T_n$ also goes down with every transition in the cyclic process L→H→L→H... if no precaution is taken to change the temperature slowly. As the number of cycles increases, the crystal deteriorates. Each successive transition originates from a new defect with a lower encoded $T_n$, but not lower than some threshold $\Delta T_{n,min}$ still above $T_0$.

*A nucleation site is stable, although only to some extent.* A crystal can store a nucleation site, together with all the encoded nucleation information, for long time. The site can also withstand some influences such as moderate annealing, internal strains, and passage of interface. Suppose, the nucleation site is located at some point A. At the encoded temperature $T_{n,A} > T_0$ the site turns into a nucleus. If the transition L → H is completed and reversed, the interface in the H → L transition will pass through the point A. This may leave the nucleation site A unaffected in a few successive cycles, so it will remain as such in the L → H runs. Ultimately it will be destroyed and the transitions L → H will not originate at point A any more. The observation is in accord with the reappearance of the same X-ray Laue pattern for several initial times noted in a long cyclical processes.

*$\Delta T_n$ in a solid → solid phase transition is smaller than in a liquid → solid phase transition.* At least, it was so in our experiments reproduced several times with different crystals. First, $\Delta T_n = |T_n - T_0|$ was measured in a PDB single crystal upon L → H and H → L transitions and was found to be in 7 to 10℃ range in both cases. Then the crystal was melted on the microscopic hot stage and permitted to cool down to room temperature. No crystallization occurred, meaning that $\Delta T_n$ (liq→sol.) exceeds 32℃.

*Neither nucleation nor growth is possible in "too perfect" single crystals.* In other words, crystal defects of *two* kinds are a necessary constituents of a solid-solid phase transition. Direct evidence of that was discovered when some vapor-grown PDB single crystals were found completely incapable of changing from L to H phase. Some of these crystals (grade 5 in Fig. 4) were very thin rod-like ones, prepared by evaporation into a glass tube, and others were found on the walls of the jar where PDB was stored. Not only did not they change into the H phase upon heating, but it was also impossible to induce this change by means of the "artificial nucleation". Even after some of the crystals were badly damaged by multiple pricks with a needle, they still "refused" to change into the H phase at the temperature as high as only 0.2℃ below $T_m$. While all conditions for *starting* nucleation were thus provided, the nuclei could not grow. Evidently, some additional condition was absent. That condition is a *sufficient concentration* of another type defects (see below). Anyhow, infeasibility of a homogeneous nucleation manifested itself quite unambiguously.

## 4. Formation of a Nucleus: A Predetermined Act, Rather Than a Successful Random Fluctuation

The conventional theory of nucleation in solids [49] considered nucleation of new phase to be a consequence of random heterophase fluctuations that give rise to formation of clusters large enough to become stable. The only change made in the formulae over nucleation in liquids and gases is an extra term representing the idea that nucleation barrier in solids is higher due to arising strains. The theory in question originally considered nucleation to be homogeneous (assuming equal probability for nuclei to appear at any point in the crystal), but had later to acknowledge that heterogeneous, *i.e.* localized at crystal defects, nucleation prevails. Therefore the theory was modified to take this into account. Heterogeneous nucleation is believed to occur at dislocation lines, foreign molecules and vacancies being present in real crystals in great numbers. In other respects the statistical-fluctuation approach has been left intact.

The real nucleation in a crystal is quite different. The observation of $\Delta T_n$ (solid→solid) < $\Delta T_n$ (liquid→solid) runs counter to the theoretical premise that activation energy of nucleation in solids is higher due to the internal strains. Formation of a nucleus is a rare and reproducible act bound to a predetermined location. Homogeneous nucleation is impossible. As for heterogeneous nucleation, the formation of nuclei at dislocation lines or vacancies has to be ruled out as inconsistent with evidence. There are plenty of these defects in every real crystal and they change their position under the action of even the slightest internal strains, to say nothing of the strains caused by moving interfaces. Dislocation lines, vacancies, and foreign molecules cannot account for the observed *rarity, stability, and reproducibility* of a nucleus. They are too primitive to contain encoded individual information about $T_n$ and orientation of the resultant crystal.

## 5. The Structure of a Nucleation Site

Finding the particular structure of a crystal defect serving as a nucleation site thus turned out to be the key to nucleation in solids. The solution is presented below. It inevitably involves an element of speculation, but only to link all elements of comprehensive evidence into a self-consistent and clear picture. It is qualitative, but more reliable than a detailed mathematical description of an idea that has not been verified.

Let us sum up the properties of these defects:

− there are few such defects in a good single crystal,

− they reside at permanent locations,

− they do not form spontaneously over long-term storage at any level of overheating / overcooling,

− they are stable enough to withstand rather strong influences,

− they are not quite as stable as macroscopic defects, − they possess a memory large enough to contain individual information both on $T_n$ and nucleus orientation

− they are capable of activating the stored information repeatedly,

− their structure permits nucleation without development of prohibitive strains.

Microcavities (cluster of vacancies) of some optimum size (Fig. 6) are, perhaps, the only type of crystal defect that meets all these requirements. An *optimum microcavity* (OM) eliminates the problem of the great strains that, possibly, prevent nucleation in a defect-free crystal medium. The OM consists of many individual vacancies and is therefore relatively stable and bound to a permanent location. There can only be a few such large-sized defects in a good single crystal, or even none. Yet, the defect in question is far from a macroscopic size, so it can be affected in some way to lose its nucleation function. Variations of its size and shape account for its capability of storing individual information on nucleation. (One additional vacancy can be attached to OM in many different ways, thus adding many new variants of "encoded" information).

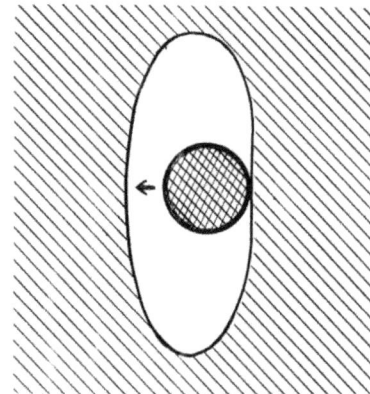

**Figure 6.** Formation of a nucleus in an optimum microcavity. Not only there are no accompanied strains, but the activation energy of molecular relocation is especially low owing to attractive action of the opposite wall

The relatively large size of OM has already been discussed. But OM also cannot be too large. Limited stability of OM is one argument in favor of such a conclusion. Now let us return to Fig. 6. It is true that a very large cavity eliminates the associated strains in the same way a smaller one does. In both cases a nucleus can grow freely on a cavity wall. The fact is, however, that a large cavity is equivalent to an external surface where no nucleation strains would be involved. But crystal faces do not facilitate nucleation, as it follows from the observation that nucleation in a "too perfect" crystal fails to occur both in the bulk and on the faces. Nucleation on the crystal faces obviously lacks some additional condition.

The idea on the nature of the lacking condition comes from the mode of molecular rearrangement at the contact interface (Sec. 2). There, molecular relocation from one side of interface to the opposite one proceeded under the attractive action of the latter, and this circumstance lowered the activation energy. Similarly, a strain-free nucleation with low activation energy is offered by a microcavity that is sufficiently narrow to facilitate the molecular relocation by an attractive action of the opposite wall. The gap must be of a molecular dimension. The particular width and configuration of this gap in a given OM may well be that same parameter which determines the encoded individual $T_n$. The detailed shape of OM may then be responsible for the encoded orientation of the nucleus. Thus, there is an intrinsic alliance between the molecular mechanisms of nucleation and growth. Both are based on the principle "relocation under attraction". When nucleus formation is completed, crystal growth takes over so naturally that these two stages of phase transition merge into a single unified process.

## 6. Epitaxial Nucleation

In general, the described nucleation site does not require structural orientation relationship. Not rarely, however, OR is observed in certain phase transitions (setting aside those where OR was incorrectly assumed). That does not mean these transitions occurred by a kind of "deformation" of the original phase or "displacement" of its molecules, as still frequently believed. They occur by nucleation and growth as well.

There are two circumstances when strict OR in phase transitions occurs. One is in layered crystal structures [8,52,53.]. A layered structure consists of strongly bounded, energetically advantageous two-dimensional units – molecular layers – usually appearing in both phases. There the interlayer interaction is weak on definition. Since the layer stacking contributes relatively little to the total lattice energy, the difference in the total free energies of the two structural variants is small. This is why layered crystals are prone to polymorphism. Change from one polymorph to the other is reduced mainly to the mode of layer stacking. The layer parameters themselves are only slightly affected by the different layer stacking.

In practice, layered structures always have numerous defects of imprecise layer stacking. Most of these defects are minute wedge-like interlayer cracks located at the crystal faces as viewed from the side of layer edges. In such a microcavity there always is a point where the gap has the optimum width for nucleation. There the molecular relocation from one wall to the other occurs with no steric hindrance and, at the same time, with the aid of attraction from the opposite wall. In view of the close structural similarity of the layers in the two polymorphs, *this nucleation is epitaxial*. Orienting effect of the substrate (the opposite wall) preserves the orientation of molecular layers.

Another case of the epitaxial nucleation is when the unit cell parameters of the polymorphs are extremely close even in non-layered crystals, as in the Fe ferromagnetic phase transition. This is also the cause of rigorous OR in case of magnetization of polydomain structures where the "polymorphs" have identical crystal structure[53,54].

The kinetics of epitaxial phase transitions differs significantly from the non-epitaxial. Hysteresis $\Delta T_n$ in epitaxial phase transitions is much smaller. Due to the abundance of the wedge-like microcracks in layered crystals, there is no shortage in the nucleation sites of optimum size. At that, the presence of a substrate of almost identical surface structure stimulates the molecular relocation. Therefore only small overheating or overcooling is required in order to initiate and quickly complete this kind of nucleation-and-growth reaction. Without a scrupulous verification, the phase transitions in question may be taken for being (kinetics-free) "displacive", "instantaneous", "cooperative", "soft-mode", "second-order", *etc.*

( *Note*: Epitaxial nucleation is the cause of formation of polydomain structures due to appearance of the nucleus in two or more equivalent positions when allowed by the substrate symmetry[53]).

## 7. Interface Motion: Additional Considerations

Section 2 outlined how the interface moves forward by a "molecule-by-molecule" relocation at the molecular steps. The availability of some extra space at the steps to provide sufficient steric freedom for the relocation was noted as an important condition. This extra space eventually comes out on the surface and a new space must appear at the interface for the process to continue. Consequently, a phase transition can take place only in a *real* crystal with a sufficient concentration of vacancies and/or their clusters. The interface kinetics described in the present section offers strong support to that basic concept. Moreover, it made possible to gain insight into the intimate details of interface motion.

We deal with two types of nuclei: 3-D (denoted as OM) to initiate a phase transition and 2-D to initiate a new molecular layer to advance the interface in the direction of its normal. The normal velocity of the interface motion $V_n$ is controlled by the latter. The 2-D nuclei form only heterogeneously, like the OM do. There is a significant difference in their function. While only one OM is needed to start a phase transition, a sufficient concentration of appropriate defects is required to keep the interface moving. These defects are also microcavities, but smaller than OM, although not just individual vacancies. They will be called *vacancy aggregates* (VAs). One VA acts as a site for the 2-D nucleus only once and then moves to the crystal surface. The VAs may differ by the number and combination of constituent vacancies, as well as how close to the interface they are.

In the process of its motion the interface intersects the positions of VAs, which is equivalent to a flow of VAs onto the interface. Intensity of the flow depends on the concentration of VAs and is affected by VAs migration. Not all VAs of the flow can be effective, but only those with the activation energy of nucleation that is lower of a particular level. That level is determined by the overheating / overcooling $\Delta T_{2\text{-}D} = |T_{tr} - T_o|$, where $T_{tr}$ is the actual temperature of phase transition. One completed phase transition in a crystal "consumes" only a part of the available VAs, leaving a possibility for the transition to be repeated..

Another major phenomenon of interface kinetics results from the fact that the interface motion is a crystallization, and, consequently, a process of purification. The cause of the purification is obvious: attachment of a proper particle to the growing crystal is more probable than of a foreign one. (Due to the "repulsion" of foreign particles by a growing crystal, crystallization from liquid phase is utilized in practice for purification of substances). In this process any crystal defects are "foreign particles" as well. In particular, vacancies and VAs form a "cloud" in front of the moving interface, as sketched in Fig. 7. This phenomenon gives rise to the following effects:

− intensification of the VAs flow to the interface, resulting in a faster interface motion;

− intensification of coagulation of vacancies into VAs and VAs into larger VAs in the "cloud";

− dissipation of the "cloud" with time due to migration of the defects towards their lower concentration, that is, away from the interface.

**Interface**

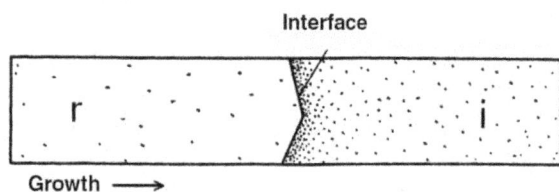

**Growth** ⟶

**Figure 7.**   Accumulation of crystal imperfections in front of moving interface. The phenomenon results from the fact that adding a proper particle to a growing crystal face is always more preferable than an improper. (Crystal defects such as vacancies or linear dislocations can also be considered "improper particles"). The well-known zone refining technique is based on the same principle

The described mechanism of interface motion is responsible only for the basic phenomena of interface kinetics. These phenomena clearly exhibit themselves under certain idealized conditions: change in the specific volume upon the phase transition of the chosen object is small, the specimens are good small single crystals, the interfaces are flat, and the velocities of the interface motion are low. If these conditions are not met, the "basic kinetics" can be completely obscured by secondary phenomena considered in Section 9.

# 8. Experimental Facts of Interface Kinetics

## 8.1. No Phase Transition in Defect-Free Crystal

This experimental fact, already described in Sec. 3, is fundamental. If crystals are "too perfect" they do not change their phase state at any temperature. Even when OMs to start phase transition are purposely created, the interface motion cannot proceed, lacking a sufficient concentration of VAs. Obviously, the crystals in which the phenomenon was observed were still far from being ideal, still containing defects such as vacancies and linear dislocations. The defects that these crystals were lacking were VAs. A new VA is required to form 2-D nucleus on the interface every time the previous one exits on the crystal surface. If the VAs are present, but their concentration is lower than required for uninterrupted 2-D nucleation, there will be no phase transition. One would be wise to take this fact into account prior to undertaking a theoretical work on phase transitions in ideal crystal medium.

## 8.2. Temperature Dependence

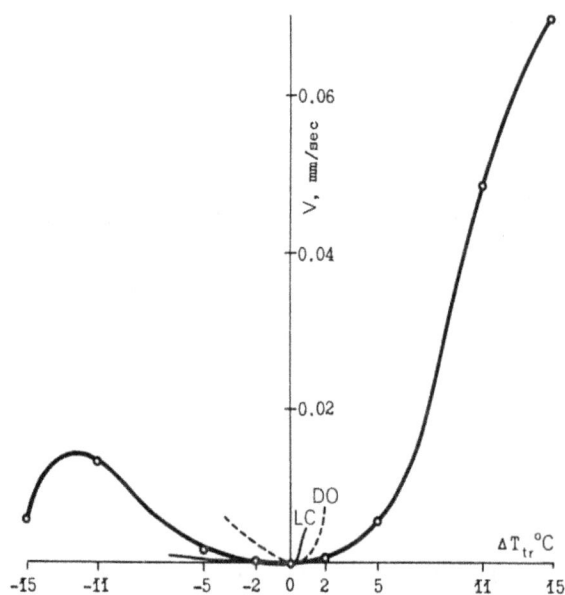

**Figure 8.**   Velocity V of interface motion in PDB against $\Delta T_{tr}$. Each experimental point was the result of averaging. The curves marked with letters were drawn from literature data: LC[*Compt. Rend.* 248, 3157 (1959)], and DO[*Docl. Acad. Nauk SSSR* 73, 1169 (1960)] to demonstrate poor reproducibility typical for kinetics measurements

There is a strong dependence of the velocity V of an interface motion on temperature. There is a problem, however, with V(T) measurements due to the V dependence on the availability of VAs. Therefore, the experimental curve in Fig. 8 should be considered as exhibiting only general qualitative features of temperature dependence. Every experimental point in the curve is the result of double averaging, first on all the transitions in each crystal, and then on different crystals. The curve shows that

♦ V = 0 at $T_o$; ($\Delta T_{tr} = T_{tr} - T_o = 0$). While $T_o$ is usually called "temperature of phase transition", it is the only temperature at which the phase transition is unconditionally impossible,

♦ V(T) is tangent to T-axis at $T_o$,

♦ V increases from zero as $T_{tr}$ moves away from $T_o$ up or down,

♦ left-hand part of the V(T) curve exhibits a maximum similar to that found in melt crystallization.

Phenomenologically, two factors shape the Fig. 8 curve: driving force and absolute temperature. The former is determined by the difference between the free energies of the phases; V is zero at $T_o$ and increases as $T_{tr}$ is moved away from $T_o$ in any direction. The absolute temperature factor, on the other hand, affects V in one direction: the higher $T_{tr}$, the higher V. To the right from $T_o$ the two factors act in the same direction, causing progressive V increase; to the left from $T_o$ they act in opposite directions giving rise to the maximum.

### 8.3. Hysteresis of Interface Motion

From the two types of temperature hysteresis, the $\Delta T_{2-D}$ to keep the interfaces moving is, as a rule, much smaller. In order to observe it, interface motion should first be stopped by setting $T_{tr} = T_o$, and then set in motion again by deviation from $T_o$. However slow and careful the last procedure is performed, one will find that some finite overheating / overcooling $\Delta T_{tr} \neq 0$ is required in order to resume interface motion. The situation in the vicinity of $T_o$ is shown schematically in Fig. 9. Interface can move only if $\Delta T_{tr}$ is greater than some threshold value. A phase transition is intrinsically a non-equilibrium phenomenon.. This conclusion is far from being trivial, considering that "non-equilibrium phase transitions" (evidently, assuming existence of the equilibrium) is a typical subject in theoretical literature.

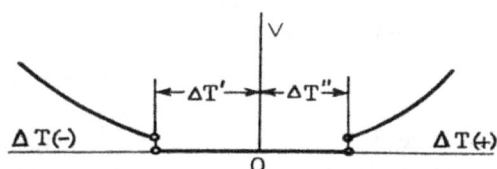

**Figure 9.** Hysteresis of interface motion. The interface motion requires 2-D nucleation, and the latter requires overcoming some energy barriers. The sketch is to illustrate the experimental fact that $\Delta T$ lower than a certain minimum ($\Delta T'$ for cooling and $\Delta T''$ for heating) will not set an interface in motion. The phenomenon is like the 3-D nucleation hysteresis, but is usually missed, or neglected, or remains unknown

### 8.4. Depletion of the Reserve of Lattice Defects

Interface motion in the small rod-like (0.22 x 5 mm) PDB single crystal shown in Fig. 10 was manipulated on a hot microscopic stage over a prolong time. The crystal, grown from a vapor phase, was of a rather high quality. By temperature control the interface (seen in the photograph) was moved back and forth many times without letting it to reach the crystal ends. After 10 to 15 cycles, the interface was moving slower in every successive cycle under the action of the same $\Delta T_{tr}$. Additional 10 to 15 cycles stabilized the interface at a fixed position: it became completely insensitive to temperature changes. In this state the specimen, consisting of the two phases divided by the interface, could

be stored for days at room temperature $20^{\circ}C$, that is, about $11^{\circ}$ lower than $T_o = 30.8^{\circ}C$. After several days the resumed experiments revealed that the dependence $V = f(\Delta T)$ had been partially restored, but the same $\Delta T_{tr}$ produced much lower V. The initial V had not been regained even after several weeks.

The schematic in Fig. 10 illustrates the cause of the phenomenon. Cyclical movements of the interface over the length $\ell$ have a "cleaning" effect. While "consuming" some part of the VAs for the 2-D nucleation, the interface pushed other VAs out of the $\ell$ area. The interface completely stopped when the concentration of VAs, $C_{VA}$, fell below the critical level required for the renewable 2-D nucleation. A partial restoration of the motion capability after the long "rest" was due to VAs migration from the end regions to the "working" region $\ell$. This experiment makes the intrinsic irreproducibility of V quite evident. The velocity in the same specimen under identical temperature conditions can differ by orders of magnitude.

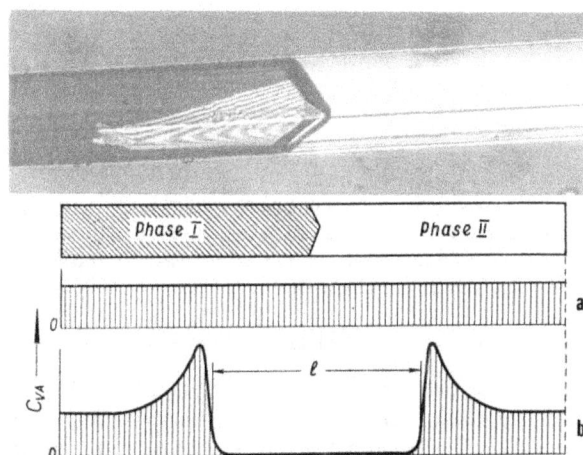

**Figure 10.** Depletion of the reserve of lattice defects (VAs) needed for 2-D nucleation to keep the interface moving. The photograph shows the thin rod-like PDB single crystal with which the experiments were carried out. The drawings show what happens to the concentration of defects $C_{VA}$, being initially uniform (plot 'a'), after the interface was moved back and forth many times over the length $\ell$ (plot 'b')

### 8.5. Velocity V as Function of the Number of Transitions

This experiment was similar to the previously described, but with two differences: the crystals were not so perfect and V was measured in every interface run. The specimens were oblong PDB single crystals grown from a solution. Due to significant V scatter the measurements were averaged over 20 crystals. A region of 1 mm long was selected in the middle of a crystal and the time required for the interface to travel this distance was measured. The outside regions played a certain auxiliary role. Two microscopes with hot stages set at $T_1 = T_o + \Delta T$ and $T_2 = T_o - \Delta T$ were used in the measurements. The V dependence on the ordinal number N of transitions in the cyclic process was measured for a fixed $|\Delta T|$. The $V = f(N)$ plots are shown in Fig. 11. Only qualitative significance should be assigned to them.

A new finding is the maxima, and more specifically, their

ascending side - because their descending side has been explained earlier. In general terms, a moving interface initially creates more VAs than it consumes, but the tendency is reversed after a number of successive transitions. In more detail, it occurs as follows. A moving interface accumulates a "cloud" of vacancies and VAs in front of it. If density of the vacancies in the "cloud" is sufficiently high, their merging into VAs creates more VAs than is expended for the 2-D nucleation. Considering that the number of vacancies in the region is limited, the consumption eventually prevails and V begins dropping. The recurrence of the whole effect after a long "rest" is due to migration of the vacancies from the end parts of the crystal.

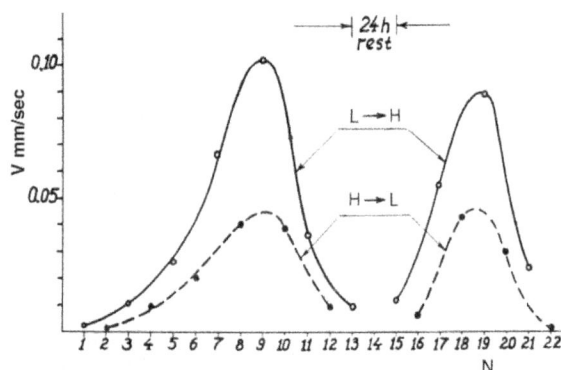

**Figure 11.** Velocity of interface motion V in PDB single crystals as a function of number N of phase transitions in the cyclic succession L→H→L→H... Two phenomena are revealed: maxima of V and ability to show them again after a sufficiently long "rest"

### 8.6. Lingering in Resting Interface Position

If an interface moving with a speed $V_1$ at a fixed $\Delta T_{tr}$ was stopped by setting $\Delta T = 0$, it tends to linger in the resting position once the initial $\Delta T_{tr}$ is restored. After some lag the interface leaves the resting position, but under a lower speed $V_2 < V_1$. The longer the resting time, the lower the $V_2$. Once resumed, the initially slow movement accelerates to approximately the previous steady $V_1$ level.

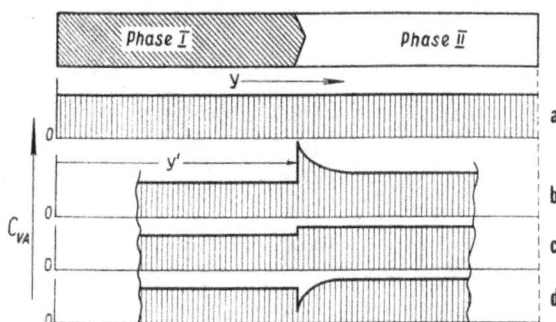

**Figure 12.** The effect of moving and resting interface on the VAs concentration, $C_{VA}$. (a) The initial uniform distribution. (b) In the course of translational motion of the interface which is in the position y' at the moment. (c) After long rest in the position y'. (d) After the motion was resumed

The diagram in Fig. 12 explains this peculiar behavior. Translational movement of interface pushes a "cloud" of

crystal defects in front of itself. Its speed $V_1$ under isothermal conditions is controlled by the density of VAs in the cloud. The density, in turn, is controlled by the balance between accumulation and consumption of VAs. Holding the interface at rest allows the cloud to dissipate to the extent depending on the resting duration. In order to start moving again, the interface must now "dig" for VAs from the uniform distribution, leaving behind a "hole". As a result, $V_2 < V_1$, but approaches $V_1$ as the new "cloud" has been accumulated.

### 8.7. Memory of the Previous Interface Position

If after the procedure just described the reverse run immediately follows, the moving interface "stumbles" (is retarded spontaneously) exactly at the position where it was previously resting. The phenomenon is almost a visual proof that the "hole" shown in Fig. 12d really exists. It provides a compelling support to the concept of interface kinetics based on the flow of VAs on the interface.

### 8.8. Slower Start upon Repetition

Using temperature control, it is possible to set up a cyclic process in which a single nucleus of H-phase will appear, grow to a certain small size, and then dissipate back to the L-phase. In such a process, growth of the H crystal in every subsequent cycle requires a longer time. Here is the cause: the growing crystal consumes the available surrounding VAs for its 2-D nucleation, while the traveling distance is too short to accumulate a "cloud" of the defects. The concentration of VAs in the area is reduced with every successive cycle, giving rise to a lover V.

### 8.9. Acceleration from Start

Just after its nucleation, an H crystal grows very slowly. It takes some traveling distance for the interface to accelerate and attain a steady V level. A "cloud" of VAs, initially absent, is then accumulated. Eventually a kind of equilibrium between their accumulation and consumption is reached, producing (in a uniform crystal medium) a translational interface motion.

### 8.10. Acceleration by Approaching Interface

When there are several H crystals growing from independent nucleation sites in the same L crystal, it can be easily seen that the rates of their growth vary in a wide range. Considering that the initial crystalline matter and $\Delta T_{tr}$ are equal for all the growing H crystals, this fact in itself is instructive in regard to kinetics of solid-state phase transitions. Which of these rates does any existing theory account for? There is another phenomenon observed repeatedly: these rates are not quite independent of one another. In one instance, pictured in Fig. 13a, the crystal $r_1$ was almost not growing when a fast-growing interface from $r_2$ began approaching from the opposite end. The latter crystal noticeably activated the growth of the former when the two were still separated by as much as 1.5 mm. As the $r_2$

was coming closer, growth of $r_1$ sharply accelerated (Fig. 13b).

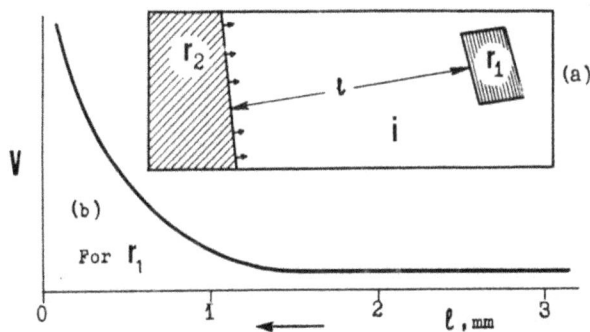

**Figure 13.** Actuation and acceleration of growth caused by an approaching interface. (a) A sketch picturing a real case when one crystal ($r_1$) was initially not growing under some overheating $\Delta T$ = const, and then was actuated by the approaching interface from $r_2$. The initial crystal is marked 'i'. (b) Change in the observed velocity V for $r_1$ (a qualitative representation, but the distances $\ell$ between $r_1$, and $r_2$ are close to real)

The crystal $r_1$ was initially not growing due to the lack of VAs in its vicinity. Transport of VAs from $r_2$ to $r_1$ has spurred its growth. The growth progressively sped up as the flow of VAs increased from the "cloud" driven by the approaching $r_2$. A plausible additional cause is the strains spreading from $r_2$ (faster-moving interfaces produce stronger strains). The strains set the static VAs dwelling at some distance from $r_1$ in motion and thus foster its growth even before it is approached by the "cloud" from $r_2$.

# 9. Revision of the Activation Energy Concept

In experimental studies of kinetics of solid-state phase transitions the phase ratio was measured *vs.* time with the objective to find the "activation energy of phase transition" $E_a$. The nature of a phase transition is heterogeneous, but $E_a$ was interpreted as the energy barrier to be overcome in the process of a cooperative homogeneous rearrangement of one ideal crystal structure into another. The inconsistence of this approach is conspicuous. Considering that phase transitions between crystal states occur by 3-D nucleation and subsequent growth, there must be at least two activation energies: one for nucleation, the other for rearrangement at the interfaces. The latter process, in turn, involves two major stages: 2-D nucleation of molecular layers and molecular relocation at interfaces. The three basic activation energies that control the above three major stages of a solid-state phase transition are:

**$E_a'$**. *Activation energy of a 3-D nucleus formation*. The $E_a'$ depends on the particular structure (size and configuration) of the lattice defect (OM) acting as the nucleation site. The nucleation temperatures encoded in these sites are different, therefore $E_a'$ is not a unique characteristic of a particular phase transition. Rather, it can be of any magnitude greater than $E'_{a,min}$ corresponding to the $\Delta T_{tr,min}$. Absence of even a

single OM in the crystal is equivalent to $E_a' = \infty$. This leads to the conclusion that attempts to find the $E_a'$ characteristic of a given phase transition would be physically unsound. This activation energy has nothing to do with interface kinetics. Phase transition in a fine-crystalline powder exemplifies the case when the bulk rate of transition under changing temperature is governed exclusively by different $E_a'$ encoded in the individual particles.

**$E_a''$**. *Activation energy of 2-D nuclei formation on a molecular-flat interface*. $E_a''$ is not a fixed value either. It varies owing to structural differences (size and shape) of VAs acting as the nucleation sites. The VAs must be present in quantities and located near the interface in order that it could be able to propagate. If this condition is not met, the phase transition (interface motion) will not be possible, which is equivalent to $E_a'' = \infty$. At moderate concentrations of VAs the interface motion is controlled more by the availability of VAs than the $E''$ magnitudes. Different speed of an interface motion at the same temperature is an example of interface kinetics governed by VAs availability. In the case of high VAs concentrations, when only a small part of the available VAs is "consumed" during interface motion, molecular relocation across the interface starts limiting the interface speed.

**$E_a'''$**. *Activation energy of molecular relocation at kinks of a contact interface*. As shown in Section 2, the process in question is a "stimulated sublimation". This activation energy is much lower then the previous two and rarely controls the linear kinetics.

# 10. Relationships between the Controlling Parameters

The complications and instabilities of interface kinetics are rooted in feedbacks. An interface needs certain conditions for its motion, but its motion affects these conditions. Flowchart below summarizes relationships between the parameters responsible for the interface kinetics controlled by VAs flow. After the foregoing discussion, the flowchart is self-explanatory even if it may seem cumbersome. Connections between the parameters should be traced from the bottom up following solid-line arrows. The feedbacks that turn the process into autocatalytic are shown by broken lines. The temperature effects are of two kinds. One is $\Delta T_{tr}$, which provides energy gradient for phase transition. The other is absolute temperature - the cause of molecular vibrations and other mobilities. The flowchart illustrates that (1) phase transition in an ideal crystal is not possible and (2) the phenomena of kinetics are complex, multiparameter and irreproducible in spite of the simplicity of the *contact* mechanism. Yet, the flowchart represents only the simplest case of slow interface motion in a good quality real single crystal and when the accompanying strains are sufficiently small not to create the additional complications described in the next section.

There is one more simplification in this flowchart, and it is essential: it does not reflect the phase transition latent heat which can dramatically affect its kinetics - up to explosion in some cases.

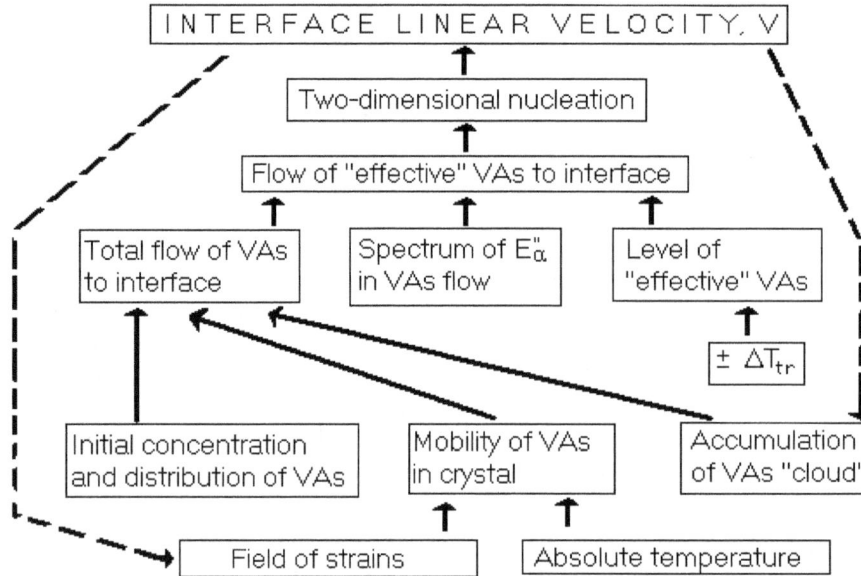

```
┌─────────────────────────────────────────────────────────┐
│ I N T E R F A C E   L I N E A R   V E L O C I T Y ,  V   │
└─────────────────────────────────────────────────────────┘
```

Flowchart:
- INTERFACE LINEAR VELOCITY, V
  - Two-dimensional nucleation
    - Flow of "effective" VAs to interface
      - Total flow of VAs to interface
      - Spectrum of $E_\alpha''$ in VAs flow
      - Level of "effective" VAs
        - $\pm \Delta T_{tr}$
- Initial concentration and distribution of VAs
- Mobility of VAs in crystal
- Accumulation of VAs "cloud"
- Field of strains
- Absolute temperature

## 11. The "Truly out of Control" Kinetics

If the previously described interface kinetics may seem "out of control", it still represents the simplest and most orderly case. A smooth advancement of a flat interface takes place only if certain precautions are taken: the specimen is a good small single crystal and $\Delta T_{tr}$ is low. It is also helpful if the specific volumes of the polymorphs are close and the crystal has a plate-like shape. Then the strains arising at the slowly moving interface can dissipate before damaging the original crystal medium. If these conditions are not favorable, the crystal growth loses visually orderly character. This disorderly morphology for a century delayed discovery of the underlying phenomenon presented in Section 2: growth of naturally-faced crystals. (Another cause of the delay was not using optical microscopy and transparent single crystals). The phase transition in most instances appears to the observer as a blurred thick "wave" rolling over the crystal and quickly completing the process, leaving behind a less transparent material. What kind of kinetics is that? The X-ray patterns reveal that a single crystal turns into a polycrystal. All facts taken together suggest that the interface *generates* multiple lattice defects, OMs, acting as the sites for 3-D nucleation immediately in front of itself. This is caused by the strains originating from the fast-moving interface. Because VAs are generated as well, the new growth proceeds quickly in both directions: toward the interface and out of it, creating new strains. Not having time for relaxation they again damage the adjacent lattice. This kinetics is based on the positive feedback:

**...interface motion → strains → generation of new nuclei in front of interface → growth from those nuclei (interface motion) → strains → ...**and so on.

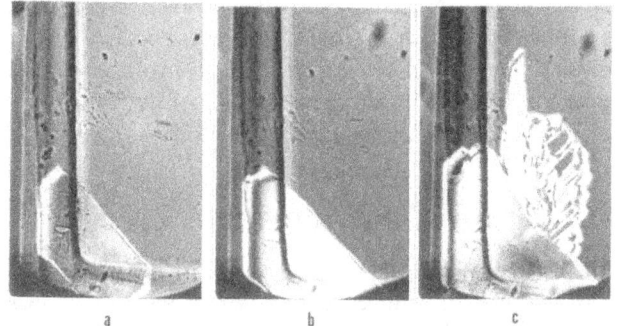

**Figure 14.** Sharp conversion of a "quiet" interface kinetics based on consumption of available defects ('a' and 'b') to that based on generation of new defects by the strains spreading from the interface ('c'). Only one corner of the original crystal is shown. The conversion manifests itself as an "explosion" on the flat surface

It should be noted that the term "interface" is used here only conditionally: actually, it is a rather thick heterophase layer. Such an interface can move with a very high speed. Fig. 14 shows an instance of a sudden change of orderly crystal growth to the kinetics based on the positive feedback. It exhibits itself as a local "explosion" with the higher speed of interface motion by one order of magnitude. Thus, in 'c' two different kinetics transparently manifest themselves in the same initial crystal.

## 12. Solid-State Recrystallization

The *contact mechanism* offers an insight into another solid-state reaction − recrystallization (migration of grain boundaries) of polycrystalline solids. It is hard to find any reason to assume one molecular mechanism for interface propagation in phase transitions and another for migration of grain boundaries. The grains in a polycrystal have the same crystal structure, but due to their random orientations the conditions at their boundaries are not different from those at phase transition interfaces. The grain boundaries do migrate. The difference is in their driving forces, namely, to minimize the grain surface energy and / or substitute a more perfect lattice for a less perfect one. They are much weaker, resulting in slower process. In all other respects it is the same crystal growth.

Recrystallization in itself is a large topic, a branch of physical metallurgy and some other applied sciences. The *contact* mechanism tells us in which direction a grain boundary moves, namely, from the grain where it has rational $(h,k,\ell)$ to the grain where it is irrational. It follows that a major component of the recrystallization driving force is the elimination of irrationally oriented boundaries. But there is more to it. The boundary will migrate only if the two neighboring grains have different orientations. The boundary between two grains of the same orientation will be either equally rational with the same $(h,k,\ell)$, or equally irrational. In such a case, no driving force to instigate the molecular relocation in one or the other direction exists: $E_a$ is same in either direction. The boundary remains still. In such a straightforward manner the *contact* mechanism accounts for one of the unexplained "recrystallization laws" which states that the boundary between two grains of the same orientation does not migrate[55].

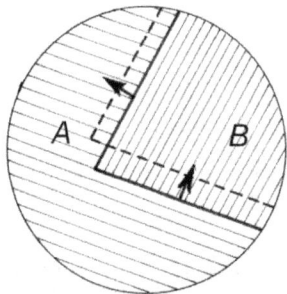

**Figure 15.** In solid-state recrystallization, the A → B and B → A migration of the boundary between grains A and B can proceed simultaneously[53]. Dashed line indicates a subsequent position of the grain boundary

One more example in the field of recrystallization. The following fact perplexed observers. Sometimes one part of the boundary between grains A and B migrates from A to B, while another part migrates from B to A (Fig. 15). In terms of the contact mechanism, the cause of the phenomenon is simply in the directions of the boundaries: in the former case, the boundary is a natural crystal plane in A and irrational in B, but the other way around in the latter case.

## 13. Ferromagnetic Transitions

Ferromagnetic phase transitions should be added to the list of solid-state reactions (and the same relates to ferroelectrics). Initially everyone believed that they are of the *second order* – a cooperative phenomenon with strictly fixed

("critical", "Curie") temperature of phase transition. In 1965 Belov wrote in his monograph "Magnetic Transitions"[56] that ferromagnetic and antiferromagnetic transitions are "concrete examples" of second-order phase transitions. But in 1970's the theorists were puzzled after a number of *first*-order ferromagnetic phase transitions were reported.. It was not realized that a first-order phase transition meant nucleation and growth, and not a critical phenomenon. Since then the number of recognized first-order ferromagnetic phase transitions grew dramatically. They turned out to be of the first order even in the basic ferromagnetics – Fe, Ni and Co. This process was accompanied by the increasing realization of structural changes involved. A new term *"magnetostructural"* transitions appeared and is being used to distinguish them from those not being " structural".

There was no explanation why some ferromagnetic phase transitions are "accompanied" by structural change, and others do not. But explanation is simple, although controversial to many (not to this author): *all ferromagnetic phase transitions are "structural"*, meaning they always materialize by nucleation and crystal rearrangements at the interfaces, rather than cooperatively. Moreover, designations of *any* phase transitions, ferromagnetic or otherwise, as second order were always superficial. Not a single sufficiently documented example exists. Ferroelectric phase transitions also materialize by nucleation and growth.

Ferromagnetic phase transition is a structural rearrangement *accompanied by* (or *giving rise to*) change of the magnetization. No change in the state of magnetization is possible without the crystal reconstruction. This is a direct consequence of the simple principle that o*rientation of a spin is imposed by the orientation of its atomic carrier*. Therefore, any reorientation of spins requires reorientation of their carriers. The only way to achieve that is replacing the crystal structure. This occurs by nucleation and interface propagation. Everything regarding the nucleation and growth in solid state (in the epitaxial version) is relevant and applicable to ferromagnetic phase transitions. *All* ferromagnetic phase transitions are "magnetostructural". The term, however, is defective in the sense that it indirectly suggests existence of ferromagnetic phase transitions without structural change. Refer to[1,15] for more details.

## 14. Magnetization by Interface Propagation

Magnetization of polydomain crystals is a solid-state reaction as well, whether the driving force is temperature, pressure or applied magnetic field. The conventional theory does not explain why magnetization is realized by propagation of interfaces rather than cooperatively in the bulk. Once again: magnetization is not a spin reorientation in the same crystal structure, it requires turning the atomic / molecular spin carriers. The only way to turn the spin carriers is by crystal rearrangement. The mechanism of crystal rearrangements is nucleation and propagation of interfaces (in this case − polydomain twin boundaries). Possibility of a cooperative magnetization "by rotation" is thus ruled out[1,15].

## 15. Conclusions

Phase transitions in solid state were studied under most refined experimental conditions. Rather high-quality small transparent single crystals were used. Perfectly-shaped single crystals of the new phase slowly grown within those crystals were observed. Both parts of the molecular mechanism of their crystal growth - nucleation and interface propagation - were experimentally examined and turned out basically the same as in crystal growth from liquids and gases. The features of that mechanism related to a solid state of the starting material were specified and became the reason for naming it "contact mechanism". The real molecular rearrangement, deduced from the experimental data, was put forward. Its nucleation part, in particular, had nothing in common with the treatment of nucleation in solids by statistical-dynamic theory. The *contact mechanism* is found to be in good agreement with available data for other reactions in solids. Besides, it is able to coherently account for the irreproducibilies and complications of kinetics of solid-state reactions.

A universal molecular mechanism of all reactions in solid state, being it a phase transition, recrystallization, magnetization, etc., is emerging. It is a *crystal growth* by nucleation encoded in the crystal defects and subsequent molecule-by-molecule relocation at the interfaces. It is universal because it is a case of crystal growth whatever the medium is - this time being a solid state. It is most energy-efficient than statistical-dynamic theories can offer, considering that it needs energy to relocate one molecule at a time rather than myriads molecules at once.

## RERERENCES

[1]    Y. Mnyukh, Fundamentals of Solid-State Phase Transitions, Ferromagnetism and Ferroelectricity, Authorhouse, 2001[or 2nd (2010) Edition].

[2]    Y. Mnyukh, Second-order phase transitions, L. Landau and his successors, Am. J. Cond. Mat. Phys., 2013, 3(2): 25-30..

[3]    Y. Mnyukh, 1963, Laws of phase transformations in a series of normal paraffins, J. Phys. Chem. Solids, 24, 631-640.

[4]    A. I. Kitaigorodskii, Y. Mnyukh, Y. Asadov, 1965, Relationships for single crystal growth during polymorphic transformation, J. Phys. Chem. Solids, 26, 463-472.

[5]    Y. Mnyukh, N. N. Petropavlov, A. I. Kitaigorodskii, 1966, Laminar growth of crystals during polymorphic transformation, Soviet Physics - Doclady, 11, 4-7.

[6]    Y. Mnyukh, N. N. Petropavlov, 1972, Polymorphic transitions in molecular crystals-1. Orientations of lattices and interfaces, J. Phys. Chem. Solids, 33, 2079-2087.

[7]    Y. Mnyukh, N. A. Panfilova, 1973, Polymorphic transitions in molecular crystals - 2. Mechanism of molecular rearrangement at 'contact' interface, J. Phys. Chem. Solids, 34, 159-170.

[8]    Y. Mnyukh, N. A. Panfilova, N. N. Petropavlov, N. S. Uchvatova, 1975, Polymorphic transitions in molecular crystals - 3. Transitions exhibiting unusual behavior, J. Phys. Chem. Solids, 36, 127-144..

[9]    Y. Mnyukh, 1976, Polymorphic transitions in crystals: nucleation, J. Crystal Growth, 32, 371-377.

[10]   Y. Mnyukh, 1979, Molecular mechanism of polymorphic transitions, Mol. Cryst. Liq. Cryst., 52, 163-200.

[11]   Y. Mnyukh, 1979, Polymorphic transitions in crystals: kinetics, Mol. Cryst. Liq. Cryst., 52, 201-218.

[12]   Y.    Mnyukh,    The    physics    of    magnetization, arxiv.org/abs/1101.1249.

[13]   Y. Mnyukh, Hysteresis and nucleation in condensed matter, arxiv.org/abs/1103.2194.

[14]   Y. Mnyukh, Lambda- and Schottky-anomalies in solid- state phase transitions, arxiv.org/abs/1104.4637

[15]   Y. Mnyukh, Ferromagnetic state and phase transitions, Am. J. Cond. Mat. Phys., 2012, 2(5): 109-115.

[16]   Y. Mnyukh, Polymorphic transitions in crystals: nucleation, J. Crystal Growth 32 (1976) 371-377.

[17]   Reactivity of Solids, Ed. J.H. de Boer, Elsevier (1961).

[18]   A.L. Mackay in Ref. 17, p. 571.

[19]   J. Hornstra in Ref. 17, p. 563.

[20]   M. Avrami, J. Chem. Phys. 7, 1103 (1939); 8, 212 (1940).

[21]   W. Kossel, Nachr. Ges. Wiss. Goetingen, 135 (1927).

[22]   I.N. Stranski, Z. Phys. Chem. 136, 259 (1928).

[23]   N.H. Hartshorne and M.N. Roberts, J. Chem. Soc., 1097 (1951).

[24]   N.H. Hartshorne, G.S. Walters and W. Williams, J. Chem. Soc., 1860 (1935).

[25]   N.H. Hartshorne, Disc. Farad. Soc. 5, 149 (1949).

[26]   W.E. Garner, Disc. Farad. Soc. 5, 194 (1949).

[27]   R.S. Bradley, N.H. Hartshorne and M. Thackray, Nature 173, 400 (1954).

[28]   R.S. Bradley, J. Phys. Chem. 60, 1347 (1956).

[29]   C. Briske and N.H. Hartshorne, Disc. Farad. Soc. 23, 196 (1957).

[30]   N.H. Hartshorne, Disc. Farad. Soc. 23, 224 (1957).

[31]   N.H. Hartshorne and M. Thackray, J. Chem. Soc., 2122 (1957).

[32]   N.H. Hartshorne, in Recent Work on the Inorganic Chemistry of Sulfur, Symp. Chem. Soc., Bristol (1958).

[33]   C. Briske, N.H. Hartshorne and D.R. Stransk, J. Chem. Soc., 1200 (1960).

[34] M. Thackray, Nature 201, 674 (1964).

[35] C. Briske and N.H. Hartshorne, Trans. Farad. Soc. 63, 1546 (1967).

[36] In Ref.1, Sections 2.4.4 and 2.4.5.

[37] Structural Phase Transitions, ed. K. Muller and H. Thomas, Springer-Verlag (1981).

[38] A.D. Bruce and R.A. Cowley, Structural Phase Transitions, Taylor and Francis (1981).

[39] Yu.M. Gufan, Structural Phase Transitions, Nauka, Moscow (1982, Rus.).

[40] R. Brout, Phase Transitions, New York, (1965).

[41] H.D. Megaw, Crystal Structures: A Working Approach, Saunders Co. (1973).

[42] N.G. Parsonage and L.A.K. Staveley, Disorder in Crystals, Clarendon Press (1978).

[43] M.E. Lines and A.M. Glass. Principles and Applications of Ferroelectrics and Related Materials, Clarendon Press (1977).

[44] Incommensurate Phases in Dielectrics, v. 2, North-Holland (1985).

[45] A.R. Verma and P. Krishna, Polymorphism and Polytypism in Crystals, Wiley (1966).

[46] Phase Transitions in Molecular Crystals, Far. Div. Chem. Soc., London (1980).

[47] Light Scattering Near Phase Transitions, Ed. H.Z.Cummins and A.P. Levanyuk, North-Holland (1983).

[48] Phase Transitions and Critical Phenomena, ed. C. Domb et al., v.1-17 (1972-1995), Acad. Press.

[49] K.C. Russell, Nucleation in Solids, in Nucleation III, Ed. A.C. Zettlemoyer, M. Dekker (1977).

[50] R. Becker, Ann. der Physik 32, 128 (1938).

[51] R.W. Cahn, in Physical Metallurgy, Ed. R.W. Cahn, North-Holland (1965).

[52] Y. Mnyukh, Phase transitions in layered crystals, arxiv.org/abs/1105.4299.

[53] In Ref. 1, Sections 2.8 and 4.2.3

[54] Y. Mnyukh, The true cause of magnetostriction, arxiv.org/abs/1103.4527

[55] W.C. McCrone, in Physics and Chemistry of the Organic Solid State, v.2, Wiley (1965).

[56] K.P. Belov, Magnetic Transitions, Boston Tech. Publ. (1965).

# Second-Order Phase Transitions, L. Landau and His Successors

There are only two ways for solid-state phase transitions to be compliant with thermodynamics: emerging of infinitesimal *quantity* of the new phase, or infinitesimal "*qualitative*" change occurring uniformly throughout the bulk at a time. The suggested theories of phase transitions are checked here for that compliance and in historical perspective. While introducing the theory of "continuous" *second-order* phase transitions, L. Landau claimed that they "may also exist" along with the majority of *first order* phase transitions, the latter being "discontinuous", displaying "jumps" of their physical properties; the fundamental differences between the two types were specified. But his theoretical successors disregarded these irreconcilable differences by presenting all phase transitions as a cooperative phenomenon treatable by statistical mechanics. In the meantime, evidence has been mounted that *all* phase transitions have a nucleation-and-growth mechanism, thus eliminating a need in the above classification.

American Journal of Condensed Matter Physics 2013, 3(2): 25-30
DOI: 10.5923/j.ajcmp.20130302.02

# Second-Order Phase Transitions, L. Landau and His Successors

**Yuri Mnyukh**

Chemistry Department and Radiation and Solid State Laboratory, New York University, New York, NY 10003, USA

**Abstract**  There are only two ways for solid-state phase transitions to be compliant with thermodynamics: emerging of infinitesimal quantity of the new phase, or infinitesimal "qualitative" change occurring uniformly throughout the bulk at a time. The suggested theories of phase transitions are checked here for that compliance and in historical perspective. While introducing the theory of "continuous" second-order phase transitions, L. Landau claimed that they "may also exist" along with the majority of first order phase transitions, the latter being "discontinuous", displaying "jumps" of their physical properties; the fundamental differences between the two types were specified. But his theoretical successors disregarded these irreconcilable differences by presenting all phase transitions as a cooperative phenomenon treatable by statistical mechanics. In the meantime, evidence has been mounted that all phase transitions have a nucleation-and-growth mechanism, thus eliminating a need in the above classification.

**Keywords**  Phase Transitions, First Order, Second Order, Landau Theory, Nucleation, Interface, Hysteresis, Ferroelectric, Magnetostructural

## 1. Compliance with Thermodynamics

Physicists in the beginning of 20th century knew that phase transitions in solid state are not "continuous" in nature. But starting from 1930's the idea of "continuous" phase transitions emerged.

When contemplating possible mechanisms of phase transitions, it should be first realized that they have, as minimum, to meet the following conditions in order to comply with thermodynamics. An infinitesimal change of a controlling parameter (dT in case of temperature) may produce only two results: either (*A*) an infinitesimal *quantity* of the new phase emerges, with the structure and properties changed by finite values, or (*B*) a physically infinitesimal "*qualitative*" change occurs uniformly throughout the whole macroscopic bulk[1]. The conditions, however, do not guarantee both versions to exist in nature.

The version '*A*' is, evidently, an abstract description of the usually observed phase transitions by nucleation and growth. Every input of a minuscule quantity of heat δQ either creates a nucleus or, if it exists, shifts the interface position by a minuscule length δℓ. The issue is whether version '*B*' can actually materialize. As far back as 1933, Ehrenfest formally classified phase transitions by *first-order* and *second-order* in terms of "continuity" or "discontinuity" in their certain

thermodynamic functions[2]. It was a theoretical exercise; the validity of the classification was disputed by Justi and Laue by asserting that there is no thermodynamic or experimental justification for second-order phase transitions[3]. Judging from the absence of references in subsequent literature, their objections were ignored.

## 2. Second-Order Phase Transitions: "May Also Exist"

Landau[4-6] developed a theory of *second-order* phase transitions. But he emphasized that transitions between different crystal modifications are "usually" *first-order*, occurring by sudden rearrangement of the crystal lattice at which the state of the matter changes abruptly, latent heat is absorbed or released, symmetries of the phases are not related and overheating or overcooling is possible. As for *second-order* phase transitions, they "may also exist", but no incontrovertible evidence of their existence was presented. It should be noted that expression that something "may exist" implicitly allows it not exist either. In case *second-order* phase transitions do exist, they must occur homogeneously, without any overheating or overcooling, at "critical points" where only the crystal symmetry changes, but structural change is infinitesimal. Landau left no doubt that his theory is that of *second-order* phase transitions only.

Since then it became accepted that there are "discontinuous" *first-order* phase transitions, exhibiting "jumps" in their physical properties, as well as "continuous" *second-order* phase transitions without "jumps". The latter

* Corresponding author:
yuri@mnyukh.com (Yuri Mnyukh)
Published online at http://journal.sapub.org/ajcmp

are to be identified with the version *'B'*, for they fit that particular version and, besides, no other option exists. Leaving alone the theory itself, there were several shortcomings in the Landau's presentation:

♦ He had not answered the arguments of the contemporaries, Max von Laue among them, that second-order phase transitions do not - and cannot - exist.

♦ The only examples he used to illustrate second-order phase transitions, $NH_4Cl$ and $BaTiO_3$, both turned out to be first order.

♦ The theory was unable to explain so called "heat capacity λ-anomalies" which, it should be noted, appeared also in first-order phase transitions.

♦ It was not specified that the only way first-order phase transitions can materialize is *nucleation and growth*;

♦ The description of first-order phase transitions left false impression that the "jump-like" changes occur simultaneously over the bulk.

♦ Overheating and overcooling in first-order transitions are not only "possible", they are inevitable (hysteresis).

♦ He remained silent when other theorists began to "further develop" his theory by treating the transitions of both types as a "critical phenomenon" in clear violation of the basic assumption of the classification in question.

## 3. First-Order Phase Transitions in More Detail

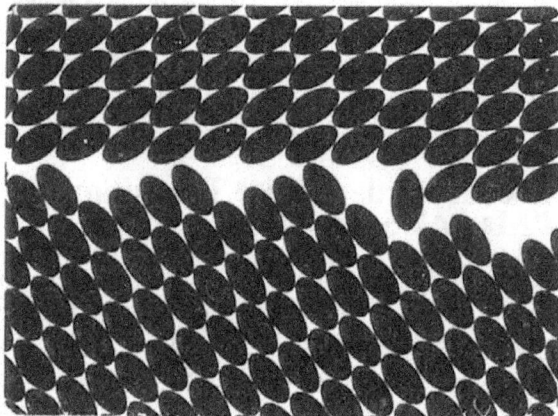

**Figure 1.** Molecular model of phase transition in a crystal. The contact interface is a rational crystal plane in the resultant phase, but not necessarily in the initial phase. The interface advancement has the edgewise mechanism. It proceeds by shuttle-like strokes of small steps (kinks), filled by molecule-by-molecule, and then layer-by-layer in this manner. (Crystal growth from liquids is realized by the same mechanism). Besides the direct contact of the two different structures, existence of the 0.5 molecular layer gap (on average) should be noted. It is wide enough to provide steric freedom for the molecular relocation at the kink, but it is narrow enough for the relocation to occur under attraction from the resultant crystal. More detailed description of the process and its advantages is given in Ref. 21 (Sec. 2.4.2-2.4.6 )

In order to better evaluate the ensuing chain of events, we need to expand Landau's characterization of first-order phase transitions by their features revealed in the subsequent studies[7-20] summarized in[21]. Solid-state phase

transitions are realized by a crystal growth involving nucleation and propagation of interfaces. Nucleation is not the classical fluctuation-based process described in the textbooks. Nucleation in a given crystal is a pre-determined process. The nuclei are located in specific crystal defects - microcavities of a certain optimum size. These defects already contain information on the condition (*e.g.,* temperature) of their activation and on orientation of the resultant crystal lattice. Nucleation lags are inevitable and reproducible for a given defect, but are not the same in different defects. The transition is an intrinsically *local* process. It proceeds by "molecule-by-molecule" structural rearrangement at interfaces only, while the bulks of the original and emerged phases remain static (Fig.1). No macroscopic "jumps" occur during the phase transition. They are simply the differences between physical properties of the initial and resultant phases, revealing themselves as "jumps" when the transition range is narrow enough.

## 4. How to Identify the 'Order'?

In order to distinguish between *first* and *second* order transitions, an indicator is needed capable to tell whether the process is local or homogeneous. The reliable indicators of first-order phase transitions are *interface, heterophase state*, and *hysteresis* - any one is sufficient, for all three are intimately linked. Thus, in principle, identification of a first-order transition is simple and definite. Not so with second-order transitions requiring proving that the above indicators are absent, while they can actually be overlooked or remain beyond the instrumental capability. The same is true for a property "jump". Its absence cannot serve as an indicator of second-order transition. Even though the participated phases are not related, the "jump" can still be tiny. The ferromagnetic phase transition in Fe at 769 °C is a good example. For decades it was regarded as the best representative of second-order phase transitions. But it was established in 2001 that it is a nucleation-and-growth phase transition, even though no "jumps" were ever reported[22]. Several years later a small latent heat - an undeniable attribute of a first-order phase transition - was recorded[23]. It is small or undetected "jumps" that were the source of erroneous classifications of phase transitions as being second-order. This method lacks the ability to tell whether the process has a local or homogeneous nature.

However small the jump is, or even looking zero, it is not an indicator of the phase transition order. Considering that a second-order phase transition is incompatible with a phase coexistence at any temperature, detection of a simultaneous presence of the two phases in any proportion at any temperature would proof the nucleation-and-growth mechanism. Presently, it can be asserted with confidence that proper verification of the remaining "second-order" phase transitions will turn them to first order. A steady process of second-to-first-order reclassification is going on. No case of reclassification in the opposite direction is known.

## 5. Blurring the Boundaries

The Landau theory initiated an avalanche of theoretical papers and books, presented not as a "theory of *second-order* phase transitions", but as a "theory of phase transitions". The first-order transitions were incorporated into a "critical phenomenon" as well. The restrictions clearly expressed by Landau that a theory of second-order transitions is not applicable to first-order ones were circumvented. Thus, Bruce and Cowley[24] avoided the "order" problem by simple replacement of the original Landau's heading[4,5] "Phase Transitions of the Second Kind" (*i.e.*, second order) by the "Landau Theory" to apply it to all phase transitions. The same road was taken by J.C. and P. Toledano[25]. Statistical mechanics was applied to many first-order transitions on the grounds that they are "almost", or "nearly", or "close to", second-order. Or, as Buerger specified, they are "90% second-order and 10% first-order"[26]. Such statement as "Although the Landau theory assumes continuous second-order phase transitions, it can be applied to weakly first-order transitions"[27] was typical. Even the very book by Landau and Lifshitz[6] had not escaped this misconception. The following footnote was placed there about $BaTiO_3$ which they used to exemplify the structural mechanism of a second-order transition: "To avoid misunderstanding it should be noted that in the particular case of $BaTiO_3$ atomic shifts experience a finite jump, although a small one, so that the transition is still that of first order". A size of the jump is irrelevant: all first-order phase transitions occur by nucleation and growth, rather than by cooperative atomic shifts.

These were examples characteristic of the whole picture. Such inseparable attributes of first-order phase transitions as nucleation, moving interfaces and a temperature range of two-phase coexistence were missing. The first-second-order classification being still recognized *de jure*, was almost abandoned *de facto*. The original intent (definitely shared by Landau) to distinguish the two antipodal types was replaced by blurring all boundaries between them in attempts to regard them as resulted from fluctuations in the bulk. The desire to treat all phase transitions as second order has turned out irresistible. The theoretical physicists wanted to apply their powerful tool - statistical mechanics. Unfortunately, it is applicable only to those solid-state phase transitions that have not yet been found.

## 6. Scaling All Solid-State Phase Transitions

Next theoretical step was the "scaling renormalization group" theory of the 1970's[28,29]. Even though it was a *theory of second-order phase transitions*, this limitation soon vanished in the same way as it happened to the Landau's theory: it became simply a *theory of phase transitions*[30]. In the instances when first-order phase transitions were not ignored, they were incorporated into the new theory. As one

author claimed, "the scaling theory of critical phenomena has been successfully extended for classical first order transitions..."[31]. Taken into account the actual physical process illustrated in Fig. 1, such "extension" had no justification.

## 7. Nucleation-and-Growth Quantum Phase Transitions ?

The ensuing theoretical development was "quantum phase transitions", put forward in the last decade of 20th century[32,33]. This theory considers all solid-state phase transitions being "classical", except their special form, called "quantum", occurring at or close to $0^{\circ}$ K. The "classical" phase transitions are claimed to be continuous and fluctuation-based, with "critical points", *etc.* The "quantum" ones are a "critical phenomenon" as well, differing from the "classical" by absence of the thermal fluctuations. A problem with this theory is that "classical" phase transitions are actually nucleation-and-growth. Even Landau with his statement that phase transitions are "usually" first order was set aside when he became an obstacle. There is no reason for the transitions that occur close to $0^{\circ}$ K not to be nucleation-and-growth. More detailed analysis of the theory is given in Ref. 21 (2nd Ed., Addendum B).

The incorporation of first-order phase transitions into the theory of "quantum" phase transitions followed: once again, nucleation and crystal growth became a homogeneous process and a "critical phenomenon".

Lastly, the meaningless generalization has achieved its culmination when "scaling ideas[were applied] to quantum first order transitions"[31].

## 8. Soft-Mode, Displacive, Topological, *etc.*

To complete the picture, some independent theoretical branches should also be mentioned, all disregarding the real nucleation-and-growth mechanism. They are: *soft-mode* concept, *displacive* phase transitions, and *topological* phase transitions.

The *soft-mode* concept[34-37] claims phase transitions to occur by sudden cooperative "distortion" of the initial phase as soon as one of the low-frequency optical modes "softens" enough toward the transition temperature. Hear we deal with the cooperative macroscopic changes not permitted by thermodynamics (conditions *'A' and 'B'* in section 1 above). More details on this subject can be found in Ref. 21 (Sec. 1.6, 1.7).

The *displacive* phase transitions were rather an idea then a theory, and no experimental proof of that idea ever existed. They were assumed from comparisons of the initial and final structures when they "looked similar". The idea was put forward by Buerger in the 1950's[26] as deformation / distortion of the original structure by cooperative displacements of the atoms/molecules in the crystal lattice without breaking their chemical bonding. It did not work well, since some bonding still had to be broken. Nevertheless, it is presently sufficient for a phase transition to be called *displacive* if the two crystal phases are "sufficiently similar". If they are not, an imaginary trajectory is constructed to achieve the transformation in several intermediate "displacive" steps. In such a case the phase transition is called *topological*. These two imaginary mechanisms cannot materialize on the same reason: they are *cooperative macroscopic jumps*. Besides, they are not needed, considering that phase transitions can (and do) occur by nucleation and growth. More on these two types are given in Ref. 21 (2nd Ed., Addendum C).

## 9. Searching for Truly Second-Order Phase Transitions

Landau himself was unable to produce a correct example of structural second-order phase transition, and no one filled the void since. Ascribing a second order to structural phase transitions is still not rare, but it is always superficial, being a side product in the investigation of something else. Not observing of hysteresis or of a large "jump" in the recording property, or taking the latent heat for heat capacity, was the "criterion". A detailed experimental investigation of a few cases seemingly lacking hysteresis and reminiscent to be second order[17] revealed that the crystal structure was layered, the hysteresis, though small, existed, and the transition proceeded by interface propagation.

The rotational order-disorder phase transitions are another instructive example. *Orientation-Disordered Crystals* (ODC) are a mesomorphic state in which the constituent particles are engaged in thermal hindered rotation, while retaining a 3-D translation crystal order. It seemed a common sense to claim that the CRYSTAL - ODC phase transitions are of second order. But the hope that second-order phase transitions found at last an ideal subject of their existence quickly faded. Landau and Lifshitz[6] warned: "There are statements in literature about a connection between second-order phase transitions and emerging rotating molecules in the crystal. This belief is erroneous..." After that it still took years for the problem to become settled. It was investigated in Ref. 21 (Sec. 2.7) and shown that such representative candidates for second-order CRYSTAL - ODC phase transition as $CBr_4$, $C_2Cl_6$, $CH_4$, $NH_4Cl$, $CBr_4$ - are realized by nucleation and growth. In the case of $C_2Cl_6$ the photographic pictures were taken[12] exhibiting growing faceted orientation-disordered crystals in the "normal" non-rotational crystal phase. The "disordering" proceeded by nucleation and crystal growth.

From 1970's some theorists abandoned looking for a good example of structural second-order phase transition and turned to ferromagnetic phase transitions[38]. Vonsovskii [39] stated that the theory of second-order phase transitions provided an "impetus" to studies of magnetic phase transitions. In view of the incessantly shrinking availability of second-order phase transitions, ferromagnetic transitions became the most reliable example of their existence, and first of all, the ferromagnetic phase transition in Fe. In 1965 Belov[40] wrote that ferromagnetic and antiferromagnetic transitions are "concrete examples" of second-order phase transitions. His work was devoted to spontaneous magnetization and other properties of Ni in the vicinity of the Curie points. The problem was, however, how to extract these "points" from the experimental data which were always "smeared out" and had "tails" on the temperature scale, even in single crystals. Unfortunately for this and other authors, they were actually dealing with all the effects that accompany first-order nucleation-growth phase transitions, namely, the temperature ranges of phase transitions and related pseudo-anomalies.

Just a few years later it was recognized that some ferromagnetic phase transitions were of the first order. In the book on magnetism by Vonsovskii[39] about 25 such phase transitions were already listed. They were interpreted in the usual narrow-formal manner as those exhibiting "abrupt" changes and / or hysteresis of the magnetization and other properties. A puzzling fact of their existence led to theoretical and experimental studies. It was always assumed that magnetization was the cause of phase transitions, while changes in the crystal parameters, density, heat capacity, etc.- the accompanying effects. The idea that change in the state of magnetization is *caused* by change in the crystal structure has not emerged. The conventional theory was in a predicament: the Curie point was not a point any more, and was rather a range of points and, even worse, was a subject to temperature hysteresis. It was not realized that a first-order phase transition meant nucleation and growth, and not a critical phenomenon. The problem of the first-order ferromagnetic phase transitions had not been resolved.

The thermodynamic theory that treats ferromagnetic phase transitions as being continuous lost its grounds. It cannot be applied even to such basic ferromagnets as Fe, Ni and Co. A "discontinuity" of the Mössbauer effect in the case of Fe was first reported in 1962 by Preston *et al.*[41], and later in more detail by Preston[42], who stated that this "might be interpreted as evidence for a first-order transition". As for Ni, the title "Mössbauer Study of Magnetic First-Order Transition in Nickel"[43] speaks for itself. The *ferromagnetic - paramagnetic* phase transition in Fe was analyzed in Ref. 21 (Sec. 4.2.3, 4.7) and concluded to be a case of nucleation and growth. Finally, the ferromagnetic phase transitions in Fe, Ni and Co were confirmed to be first order by direct experiments[23]. Yet, Fe is still used as the best example of a continuous ferromagnetic phase transition (*e.g.*,[33]). Evidently, a better example has not been found. As in case of structural phase transitions, a steady process of second-to-first-order reclassification is going on. The Google search for "first order magnetic transition", taken in January 2011 produced 2,530,000 hits, more by 20% than hits for "second order magnetic transition". Many ferromagnetic phase transitions are presently called "magnetostructural", thus assuming that there are also those not being structural. A question why some ferromagnetic transitions are combined with simultaneous structural change, and others are not, is not raised. Explanation[21 (Chapter 4), 44] of that incoherence is: *all* ferromagnetic phase transitions resulted from change of crystal structure. It is structural phase transition that brings a magnetization change about, and not the other way around.

It is presently widely accepted that "most ferroelectric phase transitions are not of second order but first[45]. "Only very few ferroelectrics...have critical or near critical transitions...the majority having first-order transitions"[46] and materialize by nucleation and growth[47]. And what about the remaining very few? That the phase transition in $BaTiO_3$ was reclassified to first order was mentioned above. The same happened to $KH_2PO_4$, even though "for years this crystal had been regarded as a typical representative of ferroelectrics undergoing second-order phase transition"[48]. The transition in TGS (tri-glycine sulfate) was believed to be the most typical second-order ferroelectric phase transition. As soon as small single-domain TGS samples were used, the characteristics of first-order phase transition were found[49]. Jumps of the electric properties and small ($\sim 0.2$ °C) hysteresis were detected[48]. The phase transition CUBIC - TETRAGONAL in $SrTiO_3$ at 105° K was confidently regarded to be second order, but later became a subject of discussion "whether pure $SrTiO_3$ possesses a first order transition or not. This question has not been clarified yet..."[50]. If the correct criteria (heterophase state, hysteresis, *etc.*) were applied to the already accumulated experimental data, its first-order mechanism would become obvious.

As happened in other cases, a second-order nature of *superconducting* phase transitions was initially taken for granted, but later became debatable. Many superconducting phase transitions has been directly named first order. The Google search for "first order superconducting phase transition", taken in January 2011, already produced 242,000 hits, more by 22% than hits for "second order superconducting phase transition". The presently available experimental data, if properly taken into account, would attest that all superconducting phase transitions are first order. They are accompanied by sharp change in some physical properties. This should not occur in second-order phase transitions by their definition.

A well-documented example of "pure" second-order superconducting transition does not exist. The claims about second order are usually based on the absence of latent heat. However, the latent heat can be small and simply avoided detection. More importantly, it has been proven[21 (Chapter 3)] that the utilized calorimetric methods of measurement do not separate *latent heat* from *heat capacity*, ascribing their combined effect to the latter. Detection of an interface, or a two-phase coexistence, or a hysteresis would proof the first-order of all those transitions. These reliable characteristics are frequently present in the experimental data, but their role as indicators of a first order not always

recognized. Some superconducting phase transitions are called "weakly first order" to treat them as second order. However, first-order phase transitions, "weakly" or not, are a local "molecule-by-molecule" process.

First-order superconducting phase transitions should have serious implications for the theories of superconductivity involving mechanism of the phase transition. The point is that all first-order phase transitions, including superconducting, are a nucleation-and-growth structural rearrangement. While comparison of the initial and resultant crystal structures may be useful, or even vital for understanding of the nature of superconductivity, the process of their crystal rearrangement is hardly specific to this kind of phase transitions.

## 10. Conclusions

Not a single sufficiently documented second-order phase transition has been found. Why? Only two mechanisms of phase transitions can comply with thermodynamics. That does not mean, however, that both will necessarily be realized. One, denoted as *first order* is the regularly observed nucleation-and-growth process. But the human-proposed *cooperative* phase transitions, denoted as *second order*, will become a reality only if they could successfully compete, at least in some cases, with the nucleation-and-growth. Then theoretical physicists would have an area for application of their talents, their knowledge of statistical mechanics and their belief in the fluctuation dominance in everything. But comparative analyses of the energy required by the two mechanisms have not been done. The important questions like *why* the phase transition in $BaTiO_3$ is of first order, and in $SrTiO_3$ is (as claimed) of second order, were not raised.

While reliable examples of second-order phase transitions have not been found, they were assumed to be a reality − with the prolonged detrimental effect to solid-state science in such areas as ferromagnetism, ferroelectricity, superconductivity, and others.

But Nature had its own agenda, namely, to make its natural processes (a) universal, (b) simple, and (c) the most energy-efficient. It produced a better process than the most brilliant human beings, even Nobel Prize winners, could invent. Solid-state phase transition by nucleation and growth, as described in Section 5, is that process. It is more universal, simple and energy-efficient than critical-dynamics theories offered. It is universal because it is just a particular manifestation of crystal growth in liquids and solids; even magnetization by magnetic field is realized by nucleation and growth[44, 51]. It is also as relatively simple as crystal growth. It is most energy-efficient because *it needs energy to relocate one molecule at a time, and not the myriads of molecules at a time* as a cooperative process requires. This is why true second-order phase transitions will never be found. The first / second-order classification is destined to be laid to rest.

# REFERENCES

[1]   M. Azbel, Preface to R. Brout, *Phase Transitions* (Russian ed.), Mir, Moscow (1967).

[2]   P. Ehrenfest, *Leiden Comm. Suppl.*, No. 75b (1933).

[3]   E. Justi and M. von Laue, *Physik Z.* 35, 945; *Z. Tech. Physik* 15, 521 (1934).

[4]   L. Landau, in *Collected Papers of L.D. Landau*, Gordon & Breach (1967), p.193.[*Phys. Z. Sowjet.* 11, 26 (1937); 11. 545 (1937)].

[5]   L. Landau and E. Lifshitz, in *Collected Papers of L.D. Landau*, Gordon & Breach (1967), p.101.[*Phys. Z. Sowiet* 8, 113 (1935)].

[6]   L.D. Landau and E.M. Lifshitz, *Statistical Physics*, Addison-Wesley (1969).

[7]   Y. Mnyukh, *J. Phys. Chem. Solids*, 24 (1963) 631.

[8]   A.I. Kitaigorodskii, Y. Mnyukh, Y. Asadov, *Soviet Physics - Doclady* 8 (1963) 127.

[9]   A.I. Kitaigorodskii, Y. Mnyukh, Y. Asadov, *J. Phys. Chem. Solids* 26 (1965) 463.

[10]  Y. Mnyukh, N.N. Petropavlov, A.I. Kitaigorodskii, *Soviet Physics - Doclady* 11 (1966) 4.

[11]  Y. Mnyukh, N.I. Musaev, A.I. Kitaigorodskii, *ibid.* 12 (1967) 409.

[12]  Y. Mnyukh, N.I. Musaev, *ibid.* 13 (1969) 630.

[13]  Y. Mnyukh, *ibid.* 16 (1972) 977.

[14]  Y. Mnyukh, N.N. Petropavlov, *J. Phys. Chem. Solids* 33 (1972) 2079.

[15]  Y. Mnyukh, N.A. Panfilova, *ibid.* 34 (1973) 159.

[16]  Y. Mnyukh, N.A. Panfilova, *Soviet Physics - Doclady* 20 (1975) 344.

[17]  Y. Mnyukh *et al.*, *J. Phys. Chem. Solids* 36 (1975) 127.

[18]  Y. Mnyukh, *J Crystal Growth* 32 (1976) 371.

[19]  Y. Mnyukh, *Mol. Cryst. Liq. Cryst.* 52 (1979) 163.

[20]  Y. Mnyukh, *ibid.*, 52 (1979) 201.

[21]  Y. Mnyukh, *Fundamentals of Solid-State Phase Transitions, Ferromagnetism and Ferroelectricity*, Authorhouse, 2001[or 2nd (2010) Edition].

[22]  Ref, 21, Sec.4.2.3 and Fig.4.2.

[23]  Sen Yang *et al.*, *Phis. Rev.* B 78,174427 (2008).

[24]  A.D. Bruce and R.A. Cowley, *Structural Phase Transitions*, Taylor and Francis (1981).

[25]  J.C. Toledano and P. Toledano, *The Landau Theory of Phase Transitions*, World Sci. (1986).

[26]  M.J. Buerger, *Kristallografiya*, 16, 1048 (1971)[*Soviet Physics - Crystallography* 16, 959 (1971)].

[27]  D.R. Moore et al., *Phys. Rev.* B 27, 7676 (1983).

[28]  K.G. Wilson, *Phys. Rev.* B 4 (1971), 3174.

[29]  K.G. Wilson,. *Scientific American*, August 1979.

[30]  M.L.A. Stile: Press Release: *The 1982 Nobel Prize in Physics*, http://nobelprize.org/nobel_prizes/physics/laureates/1982/press.html.

[31]  M.A. Continentino, cond-mat/0403274.

[32]  S. Sachdev, *Quantum Phase Transitions*, Cambridge University Press (1999).

[33]  M. Vojta, cond-mat/0309604.

[34]  W. Cochran, *Adv. Phys..* 9, 387 (1960).

[35]  *Structural Phase Transitions and Soft Modes*, Ed. E.J. Samuelson and J. Feder., Universitetsfurlaget, Norway (1971).

[36]  J. F. Scott, *Rev. Mod. Phys.* 46, 83 (1974).

[37]  G. Shirane, *Rev. Mod. Phys.* 46, 437 (1974).

[38]  H.E. Stanley, *Introduction to Phase Transitions and Critical Phenomena*, Clarendon Press (1987).

[39]  S.V. Vonsovskii, *Magnetism*, vol. 1 & 2, Wiley (1974).

[40]  K.P. Belov, *Magnetic Transitions*, Boston Tech. Publ. (1965).

[41]  R.S. Preston, S.S. Hanna and J. Heberle, *Phys. Rev.* 128, 2207 (1962).

[42]  R.S. Preston, *Phys.Rev.Let.*, 19, 75 (1967).

[43]  A.A. Hirsch, *J.Magn.Magn.Mater.* 24, 132 (1981).

[44]  Y. Mnyukh, Am. J. Cond. Mat. Phys. 2(5) (2012) 109-115.

[45]  M.E. Lines and A.M. Glass. *Principles and Applications of Ferroelectrics and Related Materials*, Clarendon Press (1977).

[46]  N.G. Parsonage and L.A.K. Staveley, *Disorder in Crystals*, Clarendon Press (1978).

[47]  *e.g.*, V.M. Ishchuk, V.L. Sobolev, *J. Appl. Phys.* 92 (2002) 2086.

[48]  I.S. Zheludev, *The Principles of Ferroelectricity*, Atomizdat, Moscow (1973, Rus.).

[49]  G.G. Leonidova, *Docl. Acad. Nauk SSSR* 196, 335 (1971).

[50]  J.O. Fossum *et al.*, *Solid State Comm.* 51 (1984), 839.

[51]  Y. Mnyukh, http://arxiv.org/abs/1101.1249.

# On the Phase Transitions that cannot Materialize

The succession of suggested mechanisms of solid-state phase transitions — Second-order, Lambda, Martensitic, Displacive, Topological, Order-Disorder, Soft-mode, Incommensurate, Scaling and Quantum — are analyzed and explained why they cannot be realized in nature. All of them assume a *cooperative* structural rearrangement as opposed to the only real one which is simply a variant of the *crystal growth*. Like all kinds of crystal growth, a solid-state phase transition proceeds by *molecule-by-molecule* building the crystal of a different structure, while the surrounding original crystal is used as the building material.

American Journal of Condensed Matter Physics 2014, 4(1): 1-12
DOI: 10.5923/j.ajcmp.20140401.01

# On the Phase Transitions that cannot Materialize

Yuri Mnyukh

76 Peggy lane, Farmington, CT 06032, USA

**Abstract**   The succession of suggested mechanisms of solid-state phase transitions − Second-order, Lambda, Martensitic, Displacive, Topological, Order-Disorder, Soft-mode, Incommensurate, Scaling and Quantum − are analyzed and explained why they cannot be realized in nature. All of them assume a *cooperative* structural rearrangement as opposed to the only real one which is simply a variant of the *crystal growth*. Like all kinds of crystal growth, a solid-state phase transition proceeds by *molecule-by-molecule* building the crystal of a different structure, while the surrounding original crystal is used as the building material.

**Keywords**   Phase transitions, First order, Second order, Lambda-transitions, Martensitic, Displacive, Topological, Order-disorder, Soft-mode, Incommensurate, Scaling theory, Quantum phase transitions

## 1. Introduction

When contemplating possible mechanisms of solid state phase transitions, a care should be taken that they would not be inconsistent with thermodynamics. An infinitesimal change of the thermodynamic parameter ($dT$ in case of temperature) may produce only two results: either (A) an infinitesimal quantity of the new phase emerges, with the structure and properties changed by finite values, or (B) a physically infinitesimal "qualitative" change occurs uniformly throughout the whole macroscopic volume[1]. These conditions, however, are only necessary ones: they do not guarantee both versions to be found in nature.

## 2. Universal Crystal Growth *vs.* Second-Order Phase Transitions

There is no doubt that version 'A' is actually realized: it is an abstract description of the usually observed phase transitions by nucleation and growth. Every input of a minuscule quantity of heat $\delta Q$ either creates a nucleus or, if it exists, shifts the interface position by a minuscule length $\delta \ell$. The issue is, however, whether version 'B' can materialize. As far back as 1933,

Ehrenfest classified phase transitions by *first-order* and *second-order*. The validity of the classification was disputed by Justi and Laue (the latter was a Noble Prize Laureate) who insisted that there is no thermodynamic or experimental justification for second-order phase transitions[2]. Landau [3,4], in disregard to those objections, developed a theory of

second-order phase transitions. Landau and Lifshitz in their book "Statistical Physics"[5] devoted a special chapter to them, claiming that they "may also exist". Since then, it became widely accepted that there are "discontinuous" *first-order* phase transitions, exhibiting "jumps" in their physical properties, as well as "continuous" *second-order* phase transitions, showing no such jumps.

The properties of the second-order phase transitions were clearly stated. Such a transition occurs at a fixed *critical* (or *Curie) point* $T_c$ where the two crystal structures are identical. There they change *continuously*; only the crystal symmetry experiences a "jump". Neither overcooling nor overheating are possible (no *hysteresis*), nor liberation or absorption of heat can take place (no *latent heat*). Coincidence of the structure orientations goes without saying. These characteristics will help in the analysis of the phase transitions that do not materialize (Sections 4 - 12). In practice, all "second-order" phase transitions fail to fit them exactly.

Prior to considering the solid-state phase transitions that do not materialize, those which do materialize should be described. They were classified as *first order* and called "usual" by Landau. He defined them as a process when the *crystal structure changes abruptly, latent heat is absorbed or released, symmetries of the phases are not related, and overheating or overcooling is possible.* In his times their molecular mechanism was not discovered yet. Later on, the systematical experimental studies by this author and associates[6-19] revealed their physical nature. The transitions were fount to be a variant of *crystal growth*, very much analogous to crystal growth from liquids or gases, but this time from a crystal medium. The results were summarized in the book[20] and articles[21-24]. Specifics of the crystal growth in a crystal medium (after a peculiar "non-classical" nucleation) is illustrated by Fig. 1.

* Corresponding author:
yuri@mnyukh.com (Yuri Mnyukh)
Published online at http://journal.sapub.org/ajcmp

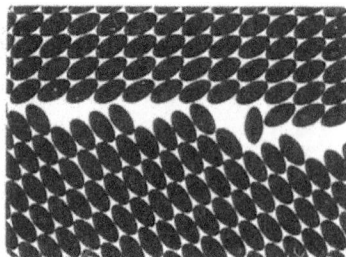

**Figure 1.** Molecular model of phase transition in a crystal. The *contact* interface is a rational crystal plane in the resultant phase, but not necessarily in the initial phase. The interface advancement has the *edgewise* mechanism. It proceeds by shuttle-like strokes of small steps (kinks), filled by molecule-by-molecule, and then layer-by-layer in this manner. The gap of 0.5 molecular layer (on average) is wide enough to provide steric freedom for the molecular relocation at the kink, but is sufficiently narrow for the relocation to occur under attraction from the side of resultant crystal

The nucleation is heterogeneous, located in *optimum microcavities*. The activation temperature $T_n$ of each potential nucleus is encoded by the microcavity size and shape. All those temperatures are different and lagging relative to the temperature point $T_o$ where the free energies of the phases are equal. *Hysteresis $\Delta T_n = T_n - T_o$ is inevitable,* and not mere possible.

An essential result of the studies was the conclusion that second-order phase transitions do not exist. All prominent examples of "second order" phase transitions turned out to be erroneous. *Justi and Laue were right when contending that there is no thermodynamic or experimental justification for second-order phase transitions.*

The remaining non-reclassified "second-order" phase transitions were usually attributed to layered crystals. Phase transitions in layer crystals have been proven[16] to materialize by nucleation and growth, but its specific morphology made it easy to assign them second-order. A

layered structure consists of strongly bounded, energetically advantageous two-dimensional units − molecular layers − appearing in both phases. There the interlayer interaction is weak on definition. Since the layer stacking contributes relatively little to the total lattice energy, the difference in the total free energies of the two structural variants is small, and so is the latent heat. Change from one polymorph to the other is reduced mainly to the mode of layer stacking. The layer parameters themselves are only slightly affected by the different layer stacking. In practice, layered structures always have numerous defects of imprecise layer stacking. Most of these defects are minute wedge-like interlayer cracks located at the crystal faces as viewed from the side of layer edges. In such a microcavity there always is a point where the gap has the optimum width for nucleation. There the molecular relocation from one wall to the other occurs with no steric hindrance and, at the same time, with the aid of attraction from the opposite wall. In view of the close structural similarity of the layers in the two polymorphs, *the nucleation is epitaxial* with a very small hysteresis. Orienting effect of the substrate (the opposite wall) preserves the orientation of molecular layers.

Now we can compare the characteristics of the *epitaxial* phase transitions with those of second-order phase transitions:

|  | **Second-order** | **Epitaxial** |
|---|---|---|
| ■ Structure orientations: | No change | layers: Same |
| ■ Structural similarity: | Identical | Very similar |
| ■ Latent heat: | Zero | Very small |
| ■ Hysteresis: | Zero | Very small |
| ■ Latent heat | Zero | Very small |

Epitaxial transition in DL-norleucine (DL-N) at ~117.2℃ [16] is an instructive example (see Fig. 2).

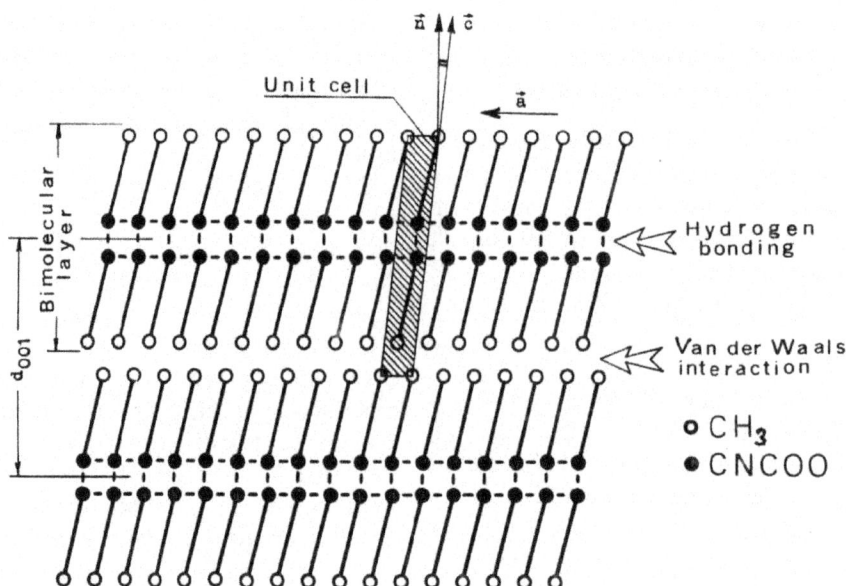

**Figure 2.** Characteristic features of the DL-*norleucine* (DL-N) crystal structure

DL-N is a short-chain aliphatic substance

$$CH_3 \cdot (CH_2)_3 \cdot CHNH_2 \cdot COOH$$

with a layered crystal structure typical of chain molecules, where the molecular axes are quite or almost perpendicular to the layer plane. Each layer is bimolecular: the *CNCOO* groups of the molecules are pointed toward the center of the layer where they form a network of hydrogen bonds *N-H...O*. This central "skeleton" turns the bimolecular layer into a firm structural unit. The interlayer interaction is much weaker, because it is of a purely Van der Waals' type, so the layer stacking is governed exclusively by the principle of close packing. As a result, both DL-N polymorphs have a pronounced layered structure of almost the same layers in different stacking mode.

Without taking special precautions, it would be easy to assign it second order: it occurs "instantly", "without hysteresis" and change of crystal orientation. But careful experimental study[16] of its single crystals (they were thin lamellae parallel to the molecular layers) has revealed: (a) It materialized by moving interfaces over the lamellae; (b) Hysteresis $\Delta T_n$ was well detectable, but was only 0.2-0.8°C; (c) The orientation of the layers did not change; (c) Laue-patterns were almost identical; (d) The layer parameters remained almost same within 1%; (e) The quantitative ratio of the coexisting phases was changing from 0% to 100% over a small temperature range; (g) The long spacing (indicator of layer stacking) changed by 4.1%.

There is a single general molecular mechanism of all solid-state phase transitions: *nucleation and crystal growth*, formerly called "first order". It exhibits itself in two forms: *epitaxial* and *non-epitaxial*. It is the former that was erroneously taken for one or another "cooperative" mechanism of phase transitions.

## 3. 300 Mechanisms of One Phenomenon

It will take a long journey before the rearrangement shown in Fig.1 is accepted as the only real molecular mechanism of solid-state phase transitions. We were able to count in the literature more than 300 types/mechanisms of solid-state phase transitions. Even if they are sorted out into groups, their number does not lend credibility to all of them; rather it indicates the failure to identify the general one. Such a state of affairs is in keen contrast with what is known about nature's laws. Nature is thrifty. There is a single equilibrium state of any solid matter, be it a metal, ionic, or organic substance: it is a *crystal state*. Crystals can come into being from vapors, melts, solutions, or other crystals. There is only one general mechanism by which crystals of any nature can emerge from any solution, vapor, or melt: it is a *nucleation and growth*. This is hardly consistent with the idea that the same process in a solid medium requires scores of diverse mechanisms.

"Transition" means a *process*: passage from one state/condition to another. Giving a name to a phase transition means an identification of the specific *mechanism*

of passage from one phase to another. This should be taken into account when looking at the collection of 300 different "mechanisms" listed in[20] (Appendix 1). Some of that chaos of names can be conditionally sorted out into groups. It is to be noted that the idea on a *cooperative* character of those mechanisms was always present, sometimes as open assumption, but mostly as a subconscious matter of course.

● Names somehow indicating at, rather than describing, the process (mechanism) of the phase transition: displacive, order-disorder, cooperative, diffusional, distortive, catastrophic, spin-flop, cation ordering, continuous... It is assumed that the phase transition is reduced to atomic/molecular displacements, structural distortion, spin-flopping, *etc*.

● Names having a more or less established theory of the mechanisms (however erroneous) in the literature: martensitic, soft mode, incommensurate, second order, quantum.

● Names carrying no characteristics at all, except being *not* something: "usual" are not martensitic, "classical" are not quantum, "structural" are not ferromagnetic, ferroelectric or superconducting, "diffusionless" are not diffusional... So are "ordinary", "normal" and "simple".

● Names of particular authors: Kastelein, Jahn-Teller, Mott, Anderson-Mott, Kosterlitz-Thouless, Berezinskii-Kosterlitz-Thouless, Ising, Lifshitz, Oguchi, Wilson, Stenley-Kaplan, Gardner, Neel, Peierls, Potts, Salam, Verway. This is a convenient way of identification: it is prestigious to those authors, absolves the responsibility to define them ...and impedes scrutiny.

● Names of the driving forces, evidently in the belief that they identify specific phase transition mechanisms: density-driven, density-driven quantum, electronically driven, driven by soft-shear acoustic mode, driven by soft mode, current-induced, pressure-induced, shock-induced, stress-induced, field-induced. That belief is invalid, considering that phase transition is driven by imbalance of free energies, and the role of any driving force is only to affect the free energy.

● A loose group of names that are too formal to reflect meaningfully on the mechanism: first order (showing "jumps" in physical properties), lambda (showing singularity of the heat capacity reminiscent to letter 'lambda'), infinite order, weak-order, non-weak, isothermal, thermodynamic, non-thermodynamic, volume-change, symmetry-breaking, symmetric-antisymmetric.

● Names indicating the prominent property of the crystal: ferromagnetic, ferroelectric, superconducting.

The unifying idea that all that diversity is the effect of a single cause − changing of the crystal structure − was missing. The following sections concentrate on those of the suggested mechanisms that significantly affected science on phase transitions and are not still completely abandoned.

## 4. Lambda-Transitions

## 4.1. Everyone Believed it is a Heat Capacity

The sharp peaks of heat capacity reminiscent to letter $\lambda$, recorded at the temperatures of solid-state phase transitions, challenged the theorists to explain their origin. The first $\lambda$-peak was observed by Simon in $NH_4Cl$ phase transition [25]. Later on, it was repeated many times and numerous other cases were reported. Thus, more than 30 experimental $\lambda$-peaks presented as "Specific heat $C_P$ of[substance] vs. temperature $T$" were shown in the book by Parsonage and Staveley[26]. The theories were unable to account for the phenomenon. P.W. Anderson wrote[27]: "Landau, just before his death, nominated[lambda-anomalies] as the most important as yet unsolved problem in theoretical physics, and many of us agreed with him... Experimental observations of singular behavior at critical points... multiplied as years went on... For instance, it have been observed that magnetization of ferromagnets and antiferromagnets appeared to vanish roughly as $(T_C-T)^{1/3}$ near the Curie point, and that the $\lambda$-point had a roughly logarithmitic specific heat $(T-T_C)^0$ nominally". Feynman stated[28] that "One of the challenges of theoretical physics today is to find an exact theoretical description of the character of the specific heat near the Curie transition - an intriguing problem which has not yet been solved."

This intriguing problem will be solved here. There were three main reasons for that theoretical impasse. (1) The $\lambda$-peaks were actually observed in first-, and not second-order phase transitions (including ferromagnetic transitions which are all "magnetostructural"[22]) (2) The first-order phase transitions exhibited latent heat, but it was mistaken for heat capacity. (3) An important limitation of the adiabatic calorimetry utilized in the measurements was unnoticed.

## 4.2. Reinterpretation of Old Experimental Data

The canonical case of "specific heat $\lambda$-anomaly" in $NH_4Cl$ around -30.6℃ will be re-examined. This case is of a special significance. It was the first where a $\lambda$-peak in specific heat measurements through a solid-state phase transition was reported and the only example used by Landau in his original articles on the theory of continuous second-order phase transitions[29]. This phase transition was a subject of numerous studies by different experimental techniques and considered most thoroughly investigated. In every calorimetric work (e.g.,[30-38]) a sharp $\lambda$-peak was recorded; neither author expressed doubts in a specific heat nature of the peak. The transition has been designated as a cooperative order-disorder phase transition of the lambda type and used to exemplify such a type of phase transitions. However, no one maintained that the $\lambda$-anomaly was understood.

It should be noted that many of the above-mentioned calorimetric studies were undertaken well after 1942 when the experimental work by Dinichert[39] was published. His work revealed that the transition in $NH_4Cl$ was spread over a temperature range where only mass fractions $m_L$ and $m_H$ of the two distinct L (low-temperature) and H

(high-temperature) coexisting phases were changing, producing "sigmoid"-shaped curves. The direct and reverse runs formed a hysteresis loop Fig. 3(a). The fact that the phase transition is first-order was incontrovertible, but not identified as such.

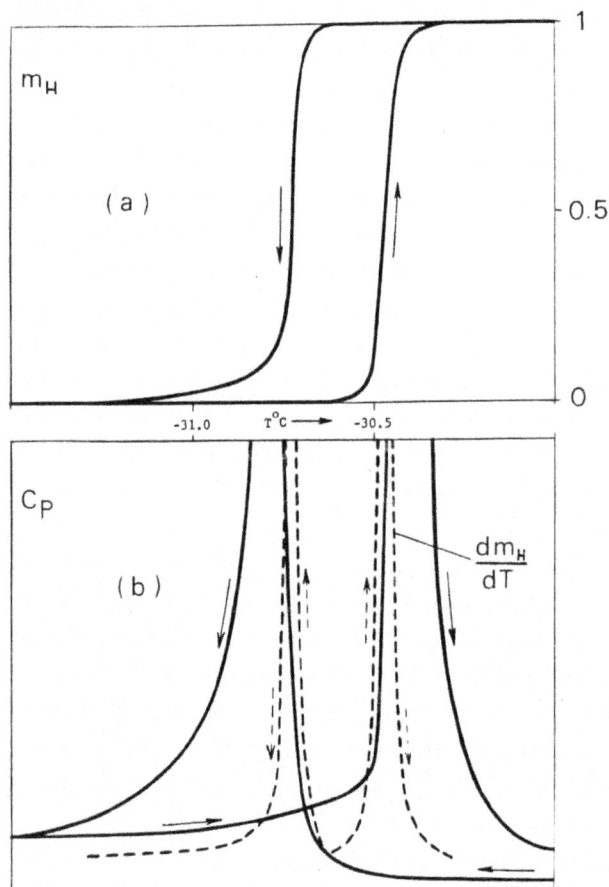

**Figure 3.** Phase transition in $NH_4Cl$. (a) The hysteresis loop by Dinichert represents mass fraction of high-temperature phase, $m_H$, in the two-phase, L+H, range of transition; $m_L+m_H = 1$. (b, solid lines) The $\lambda$-peaks from calorimetric measurements by Extermann and Weigle. The plots are positioned under one another in the same temperature scale to make it evident that the shape of the peaks is proportional to fist derivative (dotted curves) of the $m_H(T)$

In Fig. 3 the Dinichert's data are compared with the calorimetric measurements by Extermann and Weigle[32]. The latter exhibited "anomalies of heat capacity" (as the authors called the $\lambda$-peaks) and the hysteresis of the $\lambda$-peaks. Because of the hysteresis, it had already to become evident at this point (but was not) that the $\lambda$-peaks cannot be of a heat capacity, considering that heat capacity is a unique function of temperature. The graphs 'a' and 'b' are positioned under one another in the same temperature scale to reveal that the shape and location of the peaks are very close to first derivative of the $m_H(T)$ (dashed curves). It remains only to note that latent heat of the phase transition must be proportional to $dm_H/dT$. Thus, the latent heat of the first-order phase transition, lost in the numerous calorimetric studies, is found, eliminating the long-time theoretical mystery.

### 4.3. Limitations of Adiabatic Calorimetry

A legitimate question can be raised: why did not publication of the Dinichert's work change the λ-peaks interpretation from "heat capacity" to "latent heat"? The answer is: knowledge of the actual phase transition mechanism outlined in Section 2 was required. But there was also a secondary reason hidden in the calorimetric technique itself.

The goal of numerous calorimetric studies of λ-peaks in $NH_4Cl$ and other substances was to delineate shape of these peaks with the greatest possible precision. An adiabatic calorimetry, it seemed, suited best to achieve it. The adiabatic calorimeters, however, are only "one way" instruments in the sense the measurements can be carried out only as a function of increasing temperature. In the case under consideration, however, it was vital to perform both temperature-ascending and descending runs - otherwise existence of hysteresis would not be detected. And it was not detected. For example, in[37] the transition in $NH_4Cl$ was interpreted as occurring at the fixed temperature point $T_\lambda = 245.502 \pm 0.004$ K defined as a position of λ-peak. The high precision of measurements was useless: that $T_\lambda$ exceeded $T_o$ by 3°.

The results by Extermann and Weigle were not typical. The kind of calorimetry they utilized permitted both ascending and descending runs. That was a significant advantage over the adiabatic calorimetry used by others in the subsequent years. But there was also a shortcoming in their technique resulted in the unnoticed error in the presentation of the λ-peaks in Fig. 3b: the *exothermic latent heat* peak in the descending run had to be *negative* (looking downward).

### 4.4. Final Proof: It is Latent Heat

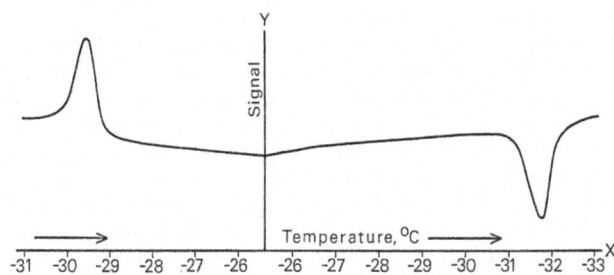

**Figure 4.** The actual DSC recording of $NH_4Cl$ phase transition cycle, displaying temperature-ascending and descending peaks as endothermic and exothermic accordingly, thus delivering final proof of a latent heat nature of the λ-peak[20] (Appendix 2)

Differential scanning calorimetry (DSC) is free of the above shortcomings[40]. Carrying out temperature descending runs with DSC is as easy as ascending runs. Most importantly, it displays endothermic and exothermic peaks with *opposite* signs in the chart recordings, which results from the manner the signal is measured[20] (Appendix 2). If the λ-peak in $NH_4Cl$ is a *latent heat* of phase transition, as was concluded above, the peak in a descending run must be

exothermic and look downward. Our strip-chart recordings made with a Perkin-Elmer DSC-1B instrument immediately revealed that the peak acquires opposite sign in the reverse run (Fig. 4). Its hysteresis was also unveiled.

## 5. Martensitic Transformations

The "martensitic" mechanism of phase transitions was one of the oldest in the succession of the proposed different mechanisms. It came from physical metallurgists who studied formation of a phase called *martensit* in iron alloys from the higher-temperature phases. This mechanism was later claimed to cover many other solid-solid phase transitions. *Martensitic transformation* was assumed to be a strictly orderly process localized at a straight interface called "habit plane". There the two crystal structures exactly match with one another, the adjacent lattices on both sides of the habit plane being under local elastic distortions to provide this matching. A martensitic transformation occurs at a specific temperature $T_M$ which is neither $T_o$, nor $T_c$. The velocity of the interface propagation is that of a sound wave, rather than a function of temperature. There must be a certain rigorous orientation relationship (OR) between the crystal lattices prior to and after the transformation. The martensitic transformation, assumed to be a cooperative at interface, was theoretically approximated by a uniform transformation in the bulk. Since direct observation of phase transitions in iron and its alloys is an extremely difficult task, the suggested "martensitic" mechanism was based more on imagination than on solid facts.

The alternative to martensitic transformations was sometimes called *diffusional*, but diffusion was a too slow process to account for the rates of "non-martensitic" phase transitions. Then the terms "usual" or "nucleation and growth" were used. These terms were not descriptive at all. There was no room for second-order phase transitions in the classification.

The more "martensitic transformations" were investigated, the more it became evident that *they do not have a single specific experimental characteristic* separating them from what was claimed to be their alternative. They start from nucleation; their actual speed was lower than that of sound propagation and depended on temperature; temperature hysteresis was not their specific feature either; OR was not always as expected, or was not strict, or was absent. All attempts to find characteristics of specifically *martensitic* mechanism have failed. They very well matched to the *nucleation and growth* as presented in Section 2.

Once dominated over a significant part of literature, the *martensitic transformations*, as a specific phase transition mechanism, was fading for a period of time until it was recently somewhat resurrected in relation to the *shape memory* effect. Now it is taken for granted; the problems with its introduction and definition are forgotten. As to the shape memory, it is actually related to the *epitaxial* phase transitions[20] (Addendum F).

## 6. Displacive Phase Transitions

The *displacive* mechanism was put forward by Buerger [41,42] solely on the basis of comparison of the crystal structures before and after a phase transition. This author shared a common belief that it was sufficient to make judgment about its *process*. A rigorous OR was assumed, but not always verified.

Buerger suggested that structures can change into one another in two ways. If they are similar, the transition does not involve breakdown of the original bonding and is *displacive*. But, if there is no way to reform the initial crystal without breaking the existing bonding net, the transition must be *reconstructive*. The descriptions given to these two mechanisms were ambiguous. The *reconstructive* transitions are first-order, but actually assumed to be cooperative. "Their structures are so different that the only way a transformation can be effected is by disintegrating one structure into small units and constructing a new edifice from the units"; such transition is "sluggish", because the substance must pass through the intermediate state of a higher energy. It suffices to note that at that time there already were plenty of experimental data on phase transitions by propagation of interfaces, the fact not being taken into account.

The description of *displacive* phase transitions was not less problematic. They are fast, barrierless, involving only a small displacement of one or more kinds of the atoms. The problem was that most, if not all, cases were "hybrids" with some bonds had to be broken. We were informed that many *displacive* transitions exhibit a small energy jump, certainly indicating first-order phase transition, but the physical rearrangement could still proceed as in second-order phase transitions. Such "firstsecond"-order hybrid phase transitions are not allowed by thermodynamics (see Section 1).

There were more drawbacks. The introduction of the two distinct types − *displacive* and *reconstructive* − turned out to be only a headline for a rather cumbersome classification. It was found impossible to relate them with the changes in the first and second coordination in the structure. Several mechanisms, such as "dilatational" and "rotational", were added. They were neither quite *displacive*, nor quite *reconstructive*. Finally, the predicted velocities of phase transitions ("rapid" or "sluggish") did not correlate with experiment. (As McCrone[43] pointed out, "one should always be ready to meet unforeseen velocities"). The whole effort was a geometrical exercise. There was no attempts to invoke thermodynamics.

If *displacive* phase transitions could exist, the DL-N (Fig. 2) could be their best example. The OR was preserved. The resultant structure could be imagined as the initial one with its rigid molecular layers simply slipped to the new mode of layer stacking. It has been proven, however, that in order to produce the almost identical new molecular layers, every original one was a subject of full molecule-by-molecule reconstruction.

## 7. Topological Phase Transitions

*Topological* phase transitions are a sophisticated version of the *displacive* ones. There are phase transitions, plenty of them, which even most inventive theorists would unable to squeeze into the "displacive" category. The mechanism of these "reconstructive" first-order phase transitions cried for explanation. The topology, a branch of mathematics, was called for help.

The "topological" approach was based on the conviction that the resultant crystal must be a *modification* of the initial one. A *cooperative continuous* character of the process was a matter of course, so there was no need to look into the experimental literature. A possibility of molecule-by-molecule reconstruction to the crystallographically independent structure did not come to mind.

So, if not by simple displacement, than how? The answer was: phase transitions proceed through several topological stages of displacements / deformations / distortions. The geometry of the participating crystal structures is analyzed and if an imaginary pass can be suggested, it is declared to be the phase transition mechanism in that particular case. Then the efforts could turn to finding the individual phase transition mechanism in next case in the same manner.

## 8. Order-Disorder Phase Transitions

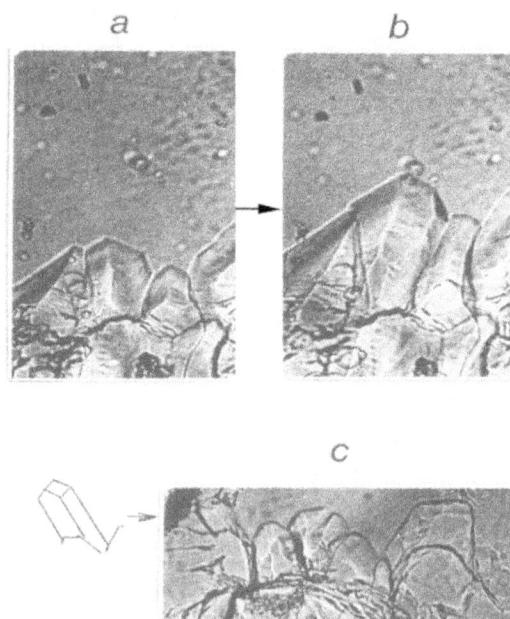

**Figure 5.** Rotational order-disorder L → H phase transition in *CBr₄*. (a,b) Growth of a conglomeration of single crystals (two successive stages). The growing ODCs are not well shaped, but the natural facing is evident. Note that the phase transition is not *cooperative*. The rotational phase (below the interface) and the non-rotational phase (above the interface) merely coexist while all phase rearrangement occurs at the interfaces. It is not a "disordering" in the bulk. (c) Another conglomeration of growing ODCs. Note the ODC reproduced in drawing

Phase transitions in which all or some constituent molecules, or their parts, of a crystal loose their definite orientations due to thermal agitation are called *order-disorder*. The resultant state was given name *orientation-disordered crystals* (ODCs). Some authors divide phase transitions into two broad types: *order-disorder* and *displacive*, implying the former to proceed by a "disordering", and the latter by "displacement", in both cases being a *cooperative* (homogeneous in the bulk) process. However, there was an important footnote in[5]: "There is claim in the literature about connection of emerging rotating molecules (or radicals) in a crystal to second-order phase transitions. That view is erroneous…". Presently the *order-disorder* phase transitions are usually assigned *first* order basically due to a noticeable density "jump", but without realization that they materialize by *nucleation and growth*. The actual crystal rearrangement in an "order-disorder" phase transition is demonstrated in Fig. 3 [11,20]. The details can be found in[20] (section 2.7).

## 9. Soft-Mode Phase Transitions

The soft-mode concept was put forward in about 1960 to explain the mechanism of *displacive* ferroelectric transitions, then applied to *order-disorder* ferroelectric transitions and, finally, tried to apply to all "structural" phase transitions. According to the developed theory, a structural phase transition is a cooperative *distortion* of the initial crystal structure as a result of atomic shifts (displacements). This distortion is produced by one of the "soft" (i.e., low-frequency) optical modes, which "softens" toward the transition temperature. When the soft-mode wavelength becomes comparable with the crystal parameters, the cooperative displacement of certain atoms makes the crystal unstable, the displacement suddenly becomes "frozen" and the crystal switches into the alternative phase. The soft-mode concept was developed, tested and demonstrated by using ferroelectric $BaTiO_3$ as an example; even "jumps" in the physical properties at the Curie point were calculated[44]. The same $BaTiO_3$ was used by Landau to illustrate a *continuous* second-order phase transition. Evidently, at least one of these conflicting approaches must be incorrect. But we will set this aside and concentrate on the soft mode model. A first-order phase transition in $BaTiO_3$ is now well established, including all the features of that phase transition type, including large hysteresis of the transition temperature.

In 1970's, the *soft-mode* theory became quite popular [44-50]). Optical and neutron spectroscopic experiments were aimed at finding a soft mode in every phase transition. In 1973 Shirane[50] distinguished two groups of phase transitions in solids: (1) magnetic and superconducting, which he regarded not being "structural" (but they are[20,22]) and maintained that they "were already reasonably understood" (but they were not at that time), and (2) "a large variety of other phase transitions", such as in $SiO_2$, $Nb_2Sn$ and those in ferroelectrics and antiferroelectrics. He contended that "the generalized soft mode concept covers the essential mechanism of phase transitions in solids" and that "the soft mode concept brings a unified picture" of how phase transitions take place in the whole second group known as *structural* phase transitions.

Not only such generalization was premature, the concept itself was not realistic.

1. It fails to comply with the minimal requirements ('A' and 'B' in Section 1) imposed by thermodynamics, considering that its instant *finite* structural "jump" at critical Curie point can only be *infinitesimal*.

2. The *instant* structural "jumps" assumed by the soft-mode concept incorrectly described the real structural phase transitions. The notion that they are instant is possibly rooted in the way Landau used the word "jumps" in describing first-order phase transitions. The irony is that the actual molecule-by-molecule rearrangement is always rather continuous. The "jump" is simply a difference in the structure and properties of the phases coexisting over a temperature range. It looks as a "jump" in the experimental measurements when the temperature range is passed quickly.

3. Being considered second-order, the soft-mode concept should not be applied even to ferroelectric phase transitions, since "only very few ferroelectrics… have critical or near critical transitions… the majority having first-order transitions"[26]; "most ferroelectric phase transitions are not of second order but first[51]". It remains to add that *all* ferroelectric phase transitions are first order and occur by nucleation and growth.

Then, how can the evidence presented in support of the soft-mode mechanism be explained? It was not definitive at all. In some cases rather "soft" modes were indeed found in the corresponding spectra of a phase, but in many other cases, including almost all molecular crystals[52], no soft modes were detected. Selection of a soft mode that "softens" toward the transition temperature was arbitrary and regarded sufficient to declare the phase transition of the soft-mode type. "Soft" modes, as any vibration modes, can be found in many crystals with or without phase transitions. Like all crystal properties, a soft mode is temperature-dependent and occasionally can show "softening" in the "desirable" direction. This in no way proves that it has any part in the phase transition, if there is one.

The soft-mode concept has not justified the hopes of its inventors. It still exists as one of the possible approaches to some solid-state phase transitions. A truly unified picture of how *all* solid-state phase transitions materialize was described in Section 2.

## 10. Incommensurate Phase Transitions

It had been well established that condensed matter can be in a liquid, crystalline, mesomorphic (liquid-crystalline or orientation-disordered-crystalline) state, or be amorphous. Then the new solid state, called *incommensurate*, was introduced and for a decade or so became very popular in certain circles of research scientists[53-57].

This new solid state was not the subject of interest *per se*.

It was invented as a remedy to cure the ailing *soft-mode* model of solid-state phase transitions. Pynn[55] asserted in the 1979 review that "the discovery and study of incommensurably distorted structures is a milestone in the investigation of structural phase transitions". In spite of the word "discovery", no evidence of the *incommensurate* state was presented in that review. As a matter of fact, no hard evidence has ever been found. Yet, the *incommensurate phase transitions* and *incommensurate* solid state were accepted as a reality.

According to the initial *soft-mode* model, a phase transition occurs under the action of a soft mode whose frequency "softens" toward the transition temperature where it turns into zero. There was a problem, however: in most real cases such an optical mode was not found. This increased doubts in the validity of the soft-mode mechanism or, at least, limited its applicability. The new idea was to "soften" requirements to the soft-mode lattice modulation. Now it did not have to "soften" further or even be a rational multiple of a dimension of the crystal unit cell. Now "the new phase does not at all possess any periodicity along the coordinate axis ...; it is referred to as incommensurate. Incommensurability may, naturally, occur along two or three coordinate axes... The fundamental feature of the crystalline state is lost"[55]. The incommensurate phase transition occurs by a "distortion" of the *underlying* ("prototype", "basic", "mother", "undistorted", "symmetrical") higher-temperature phase.

All attention in the literature was directed at the proposed new mechanism of phase transitions. No attention was paid to the resultant peculiar solid state where the displacement of every particular atom had to be unique, so that the resultant structure lacked translation symmetry. Such a solid state defies logic, our knowledge about solid state, and thermodynamics. It cannot exist for any of the following reasons.

(1) The fundamental assumption that structural phase transitions occur by a displacement (distortion, shift) is erroneous, for they occur by nucleation and growth. The relation of the soft-mode and incommensurate transitions to the first/second order classification deserved more attention than a common statement to which class one or another transition belongs. Being a *cooperative* phenomenon, they are usually regarded second-order phase transitions, but applied to first-order and "partly first-order" as well. A first-order incommensurate phase transition is an oxymoron and will not be discussed further. It cannot be of second order either: like the soft-mode transitions it should occur by a *finite* structural jump between the polymorphs and would comply neither with the second-order transitions, which are continuous, nor with thermodynamics.

(2) The theory of a *commensurate → incommensurate* transition assumes that the modulating wave becomes "frozen-in" in the resultant phase. The reverse transition could "unfreeze" it, but only with exactly the same mode. However, the vibration spectrum of the resultant phase is different and does not have that particular mode any more. Thus, the conclusion has to be drawn that this type of transition is intrinsically irreversible. What about reversible ones? The theory was silent.

(3) The polymorphs in first-order phase transitions are structurally independent, even according to Landau. But the incommensurate phase transitions assume all the lower-temperature phases of a substance to be derivatives of a "prototype" phase. Suppose there is a prototype high-temperature phase H which changes by a distortion into the lower-symmetric lower-temperature *incommensurate* phase L. The same phase L can also be obtained by growing it from a solution or vapor phase at the lower temperature where it is stable. Then we come to the absurd results: (a) the grown L crystal will have "incommensurate" rather than normal crystal structure, and (b) the grown crystal L will be a *modulated* H phase. Why does the L structure have to be "incommensurate" if the way it came into being had nothing to do with distortion of the "prototype" phase by a vibration mode? What is the source of the "intellect" that enables the crystal grown from solution to know that it must be a distorted version of another phase that can exist at a higher temperature?

(4) The alleged "incommensurate" structure cannot materialize due to a violation of the *close packing principle* valid towards metallic, ionic and molecular crystals. Violation of this principle is equivalent to rejection of the universal principle of minimum free energy in the formation of a structure. Molecular crystals are especially pictorial to illustrate the principle of close molecular packing[58]. The cause behind the principle is minimization of energy of the Van der Waals' interactions in a crystal. By encircling the molecular "skeleton" with the standard Van der Waals' radii, an organic molecule can be assigned a particular shape, as shown in Fig. 6a for biphenyl. Any real organic crystal belongs to one of the most closely packed structures of the molecules defined in this way. For an illustration, the molecular packing of the high-temperature phase of thiourea is shown in Fig. 6b.

Crystals that disobey the principle of close packing in the "incommensurate" manner are unknown. Incommensurate modulation of a prototype structure by a soft mode will cause individual molecular displacements without regard for the resultant intermolecular distances. Molecules in this structure would penetrate into one another, leaving the adjacent areas vacant. All accumulated experience to date shows that such a structure cannot exist; the polymorphs always represent two different versions of the most closely packed molecules.

To illustrate the point farther, let us turn to the mechanical model of an atomic crystal where balls represent atoms, and springs their bonds (Fig. 6c). To assume that it is possible to produce an "incommensurate" structure from this undistorted structure is equivalent to the assumption that one can displace the balls in different directions (that is, arbitrarily change the lengths of interatomic couplings in the crystal lattice) and the balls will not return to their initial equilibrium positions (i.e., the distortions will be "frozen-in", as a proponent of the incommensurate phase transition would say).

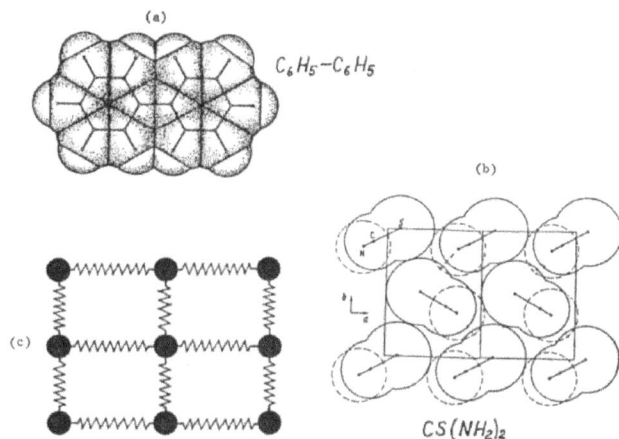

**Figure 6.** (a) The model of a biphenyl molecule constructed by encircling the molecular "skeleton" with the intermolecular radii (Kitaigorodskii[58]). (b) The close molecular packing in the high-temperature phase of thiourea. Nitrogen atoms (broken lines) are off the plane *ab* shown by solid lines. The two shown inner molecules have eight "contacts" with the surrounding neighbors (i.e. positioned at the optimum Van der Waals' distances). (c) Any irregular displacements of the balls in this model (equivalent to disturbing the network of standard interatomic distances by an "incommensurate" soft mode in an atomic crystal) will result in returning it into the shown original state. Only rearrangement leading to a new network of standard distances is plausible

Any particular "incommensurate distortion" depends on the wavelength of the mode that caused this transition ("frozen-in wave"). However, no specific mechanism of phase transition can *impose* the resultant state, because it is determined by the minimal free energy. Its position at the $p-T$ phase diagram is the exclusive function of these parameters, and not the way it arrived there. If the diagram shows the existence of two different crystal phases, the only function of the phase transition, whatever its mechanism is, is to change the above phases from one to other.

Our assertion of the "incommensurate" matter not to exist relates only to the product of the above fictitious phase transition. It does not apply to materials just because someone calls them "incommensurate", for example when some X-ray reflections are found incompatible with the lattice parameters. They resulted from the specific conditions of crystal growth, not phase transition. Thus, a phenomenon comes to mind of a "rhythmical" crystal growth from liquid phase, caused by accumulation of latent heat. Another example is "long periods" produced by folding of long-chain molecules. Such imperfect crystal structures do not violate physics of solid-state.

## 11. Scaling Mechanism of Phase Transitions

The modern theoretical physicists in the area of phase transitions pay little attention to the real solid-state phase transitions which materialize by nucleation and crystal growth over a temperature range and exhibit hysteresis.

These scientists have their own theoretical world where phase transitions are continuous / homogeneous / critical phenomenon with a fixed ("critical") point to occur and, most importantly, a subject of statistical mechanics.

Such was the "scaling renormalization group" theory of the 1970's, the subject of a Nobel Prize to K. Wilson[59,60]. Even though it was a *theory of second-order phase transitions*, this limitation soon vanished in the same way as it happened to the Landau's theory: it became simply a *theory of phase transitions*[61]. In the instances when first-order phase transitions were not ignored, they were incorporated into the new theory. As one author claimed, "the scaling theory of critical phenomena has been successfully extended for classical first order transitions…"[62]. There is no need to go into the essence of the theory in question. Whether the *scaling* theory could be fruitful in other scientific areas, it has no relation to solid-state phase transitions.

## 12. Quantum Phase Transitions

Specific "quantum" phase transitions were not the product of experimental discovery. They resulted from a theoretical idea. In order to verify legitimacy of their introduction, we turn to the review article "Quantum Phase Transitions" by M. Vojta[63]. His article is helpful on two reasons. (1) It is very authoritative, for S. Sachdev, who had published the canonical book on quantum phase transitions[64], "contributed enormously to the writing of this[Vojta's] article", and many other authorities also had "illuminating conversations and collaborations". (2) The reasons for adding the new class of phase transitions were presented in detail, which made it easier to check them for validity. Several excerptions from the Vojta's article will be used.

Excerpt: *The*[non-quantum] *phase transitions … occur at finite temperature; here macroscopic order … is destroyed by thermal fluctuations.*

That description of solid state phase transitions is imaginary. It fits to the theory of *continuous* (second-order) phase transitions, but they were not actually found and probably cannot exist at all (see Section 2). Real phase transitions are an intrinsically *local* "molecule-by-molecule" process with the bulks of the coexisting phases remaining static.

Excerpt:[Quantum phase transitions take] *place at zero temperature. A non-thermal control parameter such as pressure, magnetic field, or chemical composition, is varied to access the transition point. There, order is destroyed solely by quantum fluctuations.*

In other words, quantum phase transitions are a version of second-order phase transitions. Replacement of the thermal fluctuations by quantum is considered essential in the theory of quantum phase transitions, but leave the phenomenon to remain "continuous" and occur at "critical points". Now let us place a *real* phase transition near 0°K. The currently relocating molecule (Fig. 1) find itself in the competing attractive fields of forces emanating from the two sides of the interface. The attraction from the side of a lower free energy is stronger. Molecular vibrations, whatever they are, assist in the process, but replacement of thermal fluctuations by quantum fluctuations does not change it. The nucleation and growth will not become the subject of the quantum phase transition theory.

Excerpt:[Classical] *phase transitions are traditionally classified into first-order and continuous transitions. At first-order transitions the two phases co-exist at the transition temperature – examples are ice and water at 0 C, or water and steam at 100 C.*

To the number of different classifications of solid-state phase transitions, the "classical – quantum" was added. How "quantum" phase transitions differ from "classical"? It is not accidental that the chosen examples of first-order phase transitions were not solid-to-solid, even though "quantum" phase transitions are. The reason becomes evident since all "classical" solid-state transitions were assumed "continuous" and a "critical phenomenon". It had to be known that it is not so. It was in direct disregard of L. Landau, who is the author of the "continuous phase transitions" theory: *"Transition between different crystal modifications occurs usually by phase transition at which jump-like rearrangement of crystal lattice takes place and state of the matter changes abruptly. Along with such jump-like transitions, however, another type of transitions may also exist..."*[5]. Thus, phase transitions between crystal modifications are *first order*, but "continuous" phase transitions only *may* exist. As noted in the introduction, sufficiently documented second-order phase transitions were not found. The two phases in the *real* "classical" solid-state phase transitions coexist over a temperature range, and not only at a single temperature point.

The theory of quantum phase transitions calls all solid-state phase transitions away from 0°K "classical". Even though they are not named "second-order" in the Vojta article on some unexplained reason, they are deemed "continues" and occur at their critical points where the previously existing order is destroyed by thermal fluctuations. Toward 0°K the thermal fluctuations fade away, while the quantum fluctuations take over. The "classical" critical points become "quantum" critical points. The conclusion about existence of the "quantum" brand of phase transitions are ruined as soon as it is clarified that the "classical" phase transitions are a *nucleation and growth*. There are no critical points. The premise was erroneous.

Even though the point is now proven, it is useful to extend the analysis somewhat further.

Excerpt: *In contrast, at continuous transitions the two phases do not co-exist. An important example is the ferromagnetic transition of iron at 770 C, above which the magnetic moment vanishes. This phase transition occurs at a point where thermal fluctuations destroy the regular ordering of magnetic moments – this happens continuously in the sense that the magnetization vanishes continuously*

*when approaching the transition from below. The transition point of a continuous phase transition is also called critical point.*

Ferromagnetic phase transitions had become the last resort for the conventional theory to exemplify "continuous" phase transitions and critical phenomena. The above contradictory explanation (magnetization changes continuously at critical point) illustrates the problem to treat them as second order. It has been shown[20] (Chapter 4),[22] that they too materialize by crystal growth. As for ferromagnetic transition of *Fe*, a "discontinuity" of the Mössbauer effect there was reported already in 1962 by Preston[65,66], who stated that this "might be interpreted as evidence for a first-order transition". It was analyzed in[20] (Sec. 4.2.3, 4.7) and concluded to be a case of nucleation and growth. Finally, the first order ferromagnetic phase transitions in *Fe, Ni* and *Co* were confirmed by recording their latent heat[67].

To complete the picture, there were publications where certain "quantum" phase transitions were stated to be first order. Evidently, some authors must be incorrect. Who it was: those arguing the "quantum" phase transitions to be a "critical phenomenon" and the antithesis to first-order phase transitions, or those embracing "first-order quantum phase transitions"? The answer is: all of them are. The experimentalists, who concluded their "quantum" phase transitions being first order, are less erroneous. Their "quantum" phase transitions were first-order indeed, just not being "quantum".

# 13. Conclusions

Solid-state phase transitions were a mystery over almost all 20th century, extended to the 21st for those who do not know about the already found solutions. All that time was marked by a succession of the theories, all based on the "cooperative" idea, each one after disappointment in the previous theory. But neither theory is being completely abandoned, while the "quantum" phase transitions is still rather popular.

It is understandable how exciting it was for experimentalists to discover such anomalies as the λ–peaks, for they seemed to promise a breakthrough in a previously unexpected direction. It was not less exciting for theoretical physicists to find in the anomalies the area of application of their talents, knowledge of statistical mechanics and belief in its general power and dynamical nature of everything. But Nature had its own agenda, namely, to make its natural processes (a) universal, (b) simple and (c) the most energy–efficient. Being uncompromising in these principles, Nature produced better processes than most brilliant human beings, even Nobel Prize Laureates, could invent.

Solid-state phase transition is such a process. It is more universal, simple and energy-efficient than statistical – dynamic theories could offer. It is universal because it is just a particular manifestation of the general crystal growth. It is also as simple as crystal growth. It is energy–efficient because it needs energy to relocate one molecule at a time, and not the myriads of molecules at a time as a cooperative process requires.

An important lesson can be drawn from this. The whole effort was largely misdirected. Great amounts of time, hard work, resources and talent were wasted. Insufficient attention to facts, such as the disregard of the nucleation and growth as a mechanism inherent in all solid-state phase transitions, was substituted by excessive theoretical creativity. The contradictions were tolerated, while correct solutions were ignored. "Tries and errors" is a normal way of a scientific advancement; it is only honorable to recognize being incorrect. But that has not happened (yet?) in the area of solid-state phase transitions. As a result, the general understanding of how they materialize was unnecessarily delayed for very long time.

# REFERENCES

[1] M. Azbel, Preface to R. Brout, Phase Transitions (Russian ed.), Mir, Moscow (1967).

[2] E. Justi and M. von Laue, Physik Z. 35, 945; Z. Tech. Physik 15, 521 (1934).

[3] L. Landau, in Collected Papers of L.D. Landau, Gordon & Breach (1967), p.193.[Phys. Z. Sowjet. 11, 26 (1937); 11. 545 (1937)].

[4] L. Landau and E. Lifshitz, in Collected Papers of L.D. Landau, Gordon & Breach (1967), p.101.[Phys. Z. Sowiet 8, 113 (1935)].

[5] L.D. Landau and E.M. Lifshitz, Statistical Physics, Addison - Wesley (1969).

[6] Y. Mnyukh, J. Phys. Chem. Solids, 24 (1963) 631.

[7] A.I. Kitaigorodskii, Y. Mnyukh, Y. Asadov, Soviet Physics - Doclady 8 (1963) 127.

[8] A.I. Kitaigorodskii, Y. Mnyukh, Y. Asadov, J. Phys. Chem. Solids 26 (1965) 463.

[9] Y. Mnyukh, N.N. Petropavlov, A.I. Kitaigorodskii, Soviet Physics - Doclady 11 (1966) 4.

[10] Y. Mnyukh, N.I. Musaev, A.I. Kitaigorodskii, Soviet Physics - Doclady 12 (1967) 409.

[11] Y. Mnyukh, N.I. Musaev, Soviet Physics - Doclady 13 (1969) 630.

[12] Y. Mnyukh, Soviet Physics - Doclady 16 (1972) 977.

[13] Y. Mnyukh, N.N. Petropavlov, J. Phys. Chem. Solids 33 (1972) 2079.

[14] Y. Mnyukh, N.A. Panfilova, Soviet Physics - Doclady 34 (1973) 159.

[15] Y. Mnyukh, N.A. Panfilova, Soviet Physics - Doclady 20 (1975) 344.

[16] Y. Mnyukh et al., J. Phys. Chem. Solids 36 (1975) 127.

[17] Y. Mnyukh, J. Crystal Growth 32 (1976) 371.

[18] Y. Mnyukh, Mol. Cryst. Liq. Cryst. 52 (1979) 163.

[19] Y. Mnyukh, Mol. Cryst. Liq. Cryst. 52 (1979) 201.

[20] Y. Mnyukh, Fundamentals of Solid-State Phase Transitions, Ferromagnetism and Ferroelectricity, Authorhouse (2001);[or 2nd (2010) Edition].

[21] Y. Mnyukh, Am. J. Cond. Mat. Phys., 2013, 3(2): 25-30.

[22] Y. Mnyukh, Am. J. Cond. Mat. Phys., 2012, 2(5): 109-115.

[23] Y. Mnyukh, Am. J. Cond. Mat. Phys., 2013, 3(4): 89-103.

[24] Y. Mnyukh, Am. J. Cond. Mat. Phys., 2013,

[25] F. Simon, Ann. Phys. 68, 241 (1922).

[26] N.G. Parsonage and L.A.K. Staveley, Disorder in Crystals, Clarendon Press (1978).

[27] P.W. Anderson, Science 218, no. 4574, 763 (1982).

[28] R.P. Feynman, R.B. Leighton and M. Sands, The Feynman Lectures on Physics, v.2, Addison-Wesley (1964).

[29] L. Landau, in Collected Papers of L.D. Landau, Gordon & Breach (1967).

[30] F. Simon, C.V.Simson, and M. Ruhemann, Z. Phys. Chem. A129, 339 (1927).

[31] W.T. Ziegler and C.E. Messer, J. Am. Chem. Soc. 63, 2694 (1941).

[32] R. Extermann and J. Weigle, Helv. Phys. Acta 15, 455 (1942).

[33] V. Voronel and S.R. Garber, Sov. Phys. JETP 25, 970 (1967).

[34] W.E. Maher and W.D. McCormick, Phys. Rev. 183, 573 (1969).

[35] D.L. Connelly, J.S. Loomis, and D.E. Mapother, Phys. Rev. B 3, 924 (1971).

[36] P. Schwartz, Phys. Rev. B 4, 920 (1971).

[37] H. Chihara and M. Nakamura, Bul. Chem. Soc. Jap. 45, 133 (1972).

[38] J.E. Callanan, R.D. Weir, and L.A.K. Staveley, Proc. R. Soc. Lond. A 372, 489 (1980); 372, 497 (1980); 375, 351 (1981).

[39] P. Dinichert, Helv. Phys. Acta 15, 462 (1942).

[40] J.L. McNaughton and C.T.Mortimer, Differential Scanning Calorimetry, Perkin-Elmer, (1975).

[41] M.J. Buerger, Sov. Phys. - Crystallography 16, 959 (1971)].

[42] M.J. Buerger, in Phase Transformations in Solids, Wiley (1951).

[43] W.C. McCrone, in Physics and Chemistry of the Organic Solid State, v.2, Wiley (1965).

[44] W. Cochran, Adv. Phys. 9, 387 (1960).

[45] Structural Phase Transitions and Soft Modes, Ed. E.J. Samuelson and J. Feder, Universitetsfurlaget, Norway, (1971).

[46] C.N.R. Rao and K.J. Rao, Phase Transitions in Solids,

McGraw-Hill (1978).

[47] P.W. Anderson, in Fizika Dielectrikov, Ed. G.I. Skanavi, Akad. Nauk SSSR, Moscow (1959).

[48] R. Blinc and B. Ziks, Soft Modes in Ferroelectrics and Antiferroelectrics, North-Holland (1974).

[49] J.F. Scott, Rev. Mod. Phys. 46, 83 (1974).

[50] G. Shirane, Rev. Mod. Phys. 46, 437 (1974).

[51] M.E. Lines and A.M. Glass. Principles and Applications of Ferroelectrics and Related Materials, Clarendon Press (1977).

[52] E.R. Bernstein and B.B. Lal, Mol. Cryst. Liq. Cryst. 58, 95 (1980).

[53] Incommensurate Phases in Dielectrics, v.2, North-Holland (1985).

[54] Light Scattering Near Phase Transitions, Ed. H.Z.Cummins and A.P. Levanyuk (Ch. 3 & 7), North-Holland (1983).

[55] R. Pynn, Nature 281, 433 (1979).

[56] H. Cailleau, in Incommensurate Phases in Dielectrics, North Holland, v.2 (1985).

[57] F. Denoyer and R. Currat, in Incommensurate Phases in Dielectrics, v.2, North-Holland (1985).

[58] A.I. Kitaigorodskii, Organic Chemical Crystallography, Consultants Bureau (1960).

[59] K.G. Wilson, Phys. Rev. B 4 (1971), 3174.

[60] K.G. Wilson, Scientific American, August 1979.

[61] M.L.A. Stile: Press Release: nobelprize.org/nobel_prizes/physics/laureates/1982/press.html.

[62] M. A. Continentino, cond-mat/0403274.

[63] M. Vojta, Rep. Prog. Phys. 66 (2003) 2069-2110; cond-mat/0309604.

[64] S. Sachdev, Quantum Phase Transitions, Cambridge University Press (1999).

[65] R.S. Preston, S.S. Hanna and J. Heberle, Phys. Rev. 128, 2207 (1962).

[66] R.S. Preston, Phys. Rev. Let. 19, 75 (1967).

[67] Sen Yang et al., Phis. Rev. B 78, 174427 (2008).

# Hysteresis of Solid-State Reactions: Its Cause and Manifestations

The aim of this article is to establish the physical origin of hysteresis in solid state reactions. It had not been identified by the conventional science that remained limited by phenomenological modeling. The hysteresis revealed itself in detail as an essential component of the molecular mechanism of phase transitions in our studies of good transparent single crystals by optical microscopy and X-rays. The exclusive cause of hysteresis is *nucleation lags* and rooted in the  nature of the nucleation far from the classical "random fluctuation" model. The nuclei are localized on crystal defects where nucleation lags are encoded. The hysteresis in structural, ferromagnetic and ferroelectric phase transitions, and upon magnetization and polarization results from the underlying role of structural rearrangements in these processes. Formation of structural and magnetic hysteresis loops is analyzed in detail.

American Journal of Condensed Matter Physics 2013, 3(5): 142-150
DOI: 10.5923/j.ajcmp.20130305.05

# Hysteresis of Solid-State Reactions: Its Cause and Manifestations

Yuri Mnyukh

76 Peggy Lane, Farmington, CT, USA

**Abstract**  The aim of this article is to establish the physical origin of hysteresis in solid state reactions. It had not been identified by the conventional science that remained limited by phenomenological modeling. The hysteresis revealed itself in detail as an essential component of the molecular mechanism of phase transitions in our studies of good transparent single crystals by optical microscopy and X-rays. The exclusive cause of hysteresis is *nucleation lags* and rooted in the nature of the nucleation far from the classical "random fluctuation" model. The nuclei are localized on crystal defects where nucleation lags are encoded. The hysteresis in structural, ferromagnetic and ferroelectric phase transitions, and upon magnetization and polarization results from the underlying role of structural rearrangements in these processes. Formation of structural and magnetic hysteresis loops is analyzed in detail.

**Keywords**  Hysteresis, Structural Hysteresis, Magnetic Hysteresis, Ferroelectric Hysteresis, Phase Transition, Range of Transition, Nucleation, Hysteresis Loops, Magnetization

## 1. The Phenomenon of Hysteresis

*Hysteresis* is a lagging to respond when the system is prompted to change by a control parameter. The phenomenon does not depend on the direction of the change and is not rooted in kinetics. In solid-state reactions we will call it *structural* when it relates to change of crystal structure and *magnetic* when it relates to change of magnetization (or simply *hysteresis of polarization* in relation to electric polarization). The term "hysteresis" in the condensed matter literature is frequently assumed only the magnetic hysteresis.

Hysteresis is a significant feature of phase transitions and other solid-state reactions. The phase transitions caused by change of temperature or pressure, or by application of magnetic or electric field, occur not at the point ($T_O$ if it is temperature) when the equality of free energies of the phases is achieved, but *after* that by a finite value. In technology the phenomenon is found useful in some applications and detrimental in others, but that topic is outside the frames of the present article.

The following questions are critical when dealing with the hysteresis. Are there solid-state reactions without hysteresis? Why it is wide in some cases and narrow in others? Is its magnitude a physical constant for a given phase transition? If observed, can it be reduced? Can it be completely eliminated? If yes, then how? Why and how its notorious loops form, and

what are the factors that determine their shape? What is the relation, if any, between the temperature and ferromagnetic (and ferroelectric) hysteresis?

## 2. Conventional Science: Phenomenological Modeling

Setting aside empirical data and merely opinions, scientifically-founded answers with explanations to the above questions are not, and cannot be, found in the literature. They are hidden in the origin of the hysteresis that has not been identified.

Those who believe in the Landau's theory[1] of second-order phase transitions must also accept that those phase transitions are hysteresis-free. This inference is immaterial, however, since sufficiently documented second-order phase transitions are not found[2,3]. Then hysteresis is a feature of first-order phase transitions only. But is it inevitable in all of them? According to generally held views the transition "may or may not occur" at the temperature $T_O$ when the free energies of the two forms become equal. The actual mechanism of first-order phase transitions[2,4,5], briefly summarized in Section 4, will explain why they were incorrect at that point.

There is no lack of literature on hysteresis. The three-volume (2160 pages) set *The Science of Hysteresis*[6] published in 2007 is an example. All that science was devoted to mathematical descriptions of its manifestations. A more appropriate title would be "The Science of Hysteresis Modeling" due to absence there of anything about the physical nature of the phenomenon. From practical point of

* Corresponding author:
yuri@mnyukh.com (Yuri Mnyukh)
Published online at http://journal.sapub.org/ajcmp

view the theoretical modeling of hysteresis is useful. But

wouldn't it be better to do it while already understanding its physical origin? Probably some of that modeling would not be even needed.

The fact that the origin of hysteresis was not identified by solid state science does not mean it had not been discovered and publicly revealed. It was done already in 1979 in regard to the structural hysteresis in a special section "Hysteresis of Polymorphic Transitions" in the article *Molecular Mechanism of Polymorphic Transitions*[7]. Not one of the ensuing references to it has been related to the above section. The finding of a *nucleation* as being the exclusive cause of hysteresis is remaining unknown. The purpose of this article is to fill the void. The structural hysteresis in solids is presented here in more detail and in conjunction with the magnetic hysteresis.

## 3. "What Causes Magnetic Hysteresis?"

Probably, the most known and consequential is the *magnetic hysteresis*. Its cause has not been found by the current theory. Here the hysteresis modeling dominates as well. The regular *International Conferences on Hysteresis and Micromagnetic Modeling* (the $9^{th}$ was held in 9013) illustrate this fact.

But the sixteen authors of[8] thought it would be beneficial to understand it, pointing out that magnetic hysteresis is fundamental to magnetic storage technologies and a cornerstone to the present information age. They posed a question "What causes magnetic hysteresis?" and stated that all the "beautiful theories of magnetic hysteresis based on random microscopic disorder" failed to explain their own data. Their answer to the above question was: "New advances in our fundamental understanding of magnetic hysteresis are needed".

No such advances are possible without proper understanding of ferromagnetic state and phase transitions first. The insurmountable obstacle to that understanding is the conventional belief that ferromagnetic phase transitions are, according to the common classification, of a "second order", in spite of the fact such phase transitions must not exhibit hysteresis. (As stated by Vonsovskii[9], the theory of second-order phase transitions provided an "impetus" to studies of magnetic phase transitions). The new fundamentals of ferromagnetism and phase transitions [2,10,11] remove that basic contradiction. They demonstrate that the standard *exchange field* theory of ferromagnetism by Heisenberg, assuming existence of additional extremely strong spin interaction, has not been successful and was not even needed. It was replaced by the *crystal field* and a natural assumption that *spin orientation is uniquely bound to its carrier*. It followed at once that ferromagnetic state is *a property* of the crystal structure, that both ferromagnetic phase transitions and magnetization require structural rearrangements. The contribution of the magnetic interaction into the total free energy of a ferromagnetic crystal is

relatively small rather than dominant. The magnetic structure (the positional and orientational spatial distribution of the spin carriers and their spins in the crystal lattice) is dictated by the crystal packing, and not, as previously assumed, the other way around.

The mechanism of structural rearrangements in solids is a nucleation and interface propagation[2,4,5,7]. The *structural hysteresis* is one of its manifestations which in case of magnetic crystals is inevitably accompanied by the *magnetic hysteresis*.

It can be seen now that finding the source of magnetic hysteresis was impossible without the new understanding that magnetic phase transitions and magnetization resulted from the crystal-structural rearrangements. The answer to the above question "What causes magnetic hysteresis?" was given in the book[2] six years prior to the time the question was posed. The answer was: *nucleation*. Somehow the sixteen authors of[8] missed it.

## 4. The Cause of Structural Hysteresis as Seen in Optical Microscope

As to the cause of structural hysteresis, it will be illuminating to recall the experiments performed many years ago[12]. Phase transitions in small (~1 mm) transparent single crystals of *p*-dichlorobenzene (PDB) were investigated by direct observation in optical microscope equipped with a heating/cooling stage. Their temperature $T_0$ = 30.8℃ when the free energy $F$ of its H (above $T_0$) and L (below $T_0$) phases are equal, $F_H(T_0) = F_L(T_0)$, was convenient to do the observations. Every crystal was subjected to a slow heating or/and cooling. Fig. 1 is a photograph of phase transition in one of the heating experiments. Here are some results. Phase transitions start from nucleation *after* the $T_0$ has been passed. Nucleation is always *heterogeneous*, located at the crystal defects. The actual temperature of phase transitions did not − and could not − coincide with $T_0$, considering that no reason exists for the transition to go in any direction when $F_H = F_L$. In other words, the temperature $T_0$, usually called "phase transition temperature" (and sometimes even "critical temperature") is the temperature where phase transitions cannot occur. But the possibility to occur at any other temperatures, except of a small region around $T_0$, is theoretically unlimited. Heterogeneous nucleation requires a finite energy for activation, which makes *threshold* nucleation lags inevitable. If $T_n$ is the actual temperature of nucleation, the minimum (threshold) overheating in PDB was $\Delta T_n = T_n - T_0 = \sim 1.9℃$. The threshold overcooling was always greater. An assertion by Landau and Lifshitz[1] that overheating or overcooling is *possible* in first-order phase transitions is not applicable to solid-state reactions. There hysteresis is inherently *inevitable*. It can be small, but finite (see Section 8). In general, better crystals exhibit wider *hysteresis*. An extreme case is the observation of melting at 53.2℃ of the PDB crystals still in their L-phase, being "too perfect" to contain even a single suitable defect to serve as a nucleation site of the H-phase. This observation also illustrates absence of a homogeneous nucleation. The hysteresis in PDB crystals was typical for all other investigated objects.

**Figure 1.** An example of phase transition in a transparent single crystal of *p*-dichlorobenzene. It is a crystal growth of well-shaped single crystal of the new (higher-temperature) phase within the lower-temperature phase. The transition started from a visible crystal defect at $T_n > T_o$ by several degrees. There is no rational crystallographic orientation relationship between the initial and new phases

## 5. "Non-classical" Peculiarities of Nucleation

Thus, observations of real phase transitions of transparent single crystals bring about essential information on the origin of their hysteresis. *Nucleation* is its exclusive cause. Even more is hidden in the peculiarities of the nucleation[2,5,13]. Only *optimum microcavities* - conglomerations of vacancies, and not any other kind of defects, serve as the nucleation sites. The nucleation is not a random successful fluctuation as usually assumed, e.g.[14], it is *predetermined*. Special experiments have revealed that every potential nucleation site contains a "pre-coded" individual temperature of its activation $T_n$. In the cyclical phase transitions the $T_n$ was the same only as long as the transition was initiated in the same nucleation site. Generally, if the temperature is slowly rising, and there are several potential nucleation sites, the one of lowest temperature $T_n$ would be actually activated. The $T_n$ was different in different crystals. These specific features of nucleation of solid-state reactions provide previously unavailable authentic interpretation of the hysteresis itself and all its manifestations.

## 6. Range of Transition

The "jumps" of physical properties in phase transitions are never instant. Upon heating or cooling they always spread over a temperature range, narrow or wide, exhibiting a "sigmoid" curve (Fig. 2). It is usual to (erroneously) take the

inflection point of the curve as the "transition temperature" or even as the "critical point". Phase transitions over "wide" temperature range are called *diffuse*. The "diffuseness" is not the manifestation of a specific transition mechanism. It results from the non-simultaneous nucleation in different particles, or parts, of the specimen. Any transition in a powder or polycrystalline specimen is "diffuse", the variation being only how much. The width of a transition range is not a fixed value, being a characteristic of the particular crystal imperfection, rather than an inherent property of the substance. For instance, the range of transition will be sharply different for a single crystal and the powder made from it. Concluding, (a) range of transition is a range of *nucleation*, (b) it is affected by the sample condition / preparation, (c) it is located entirely outside $T_0$ : above it upon heating and below it upon cooling.

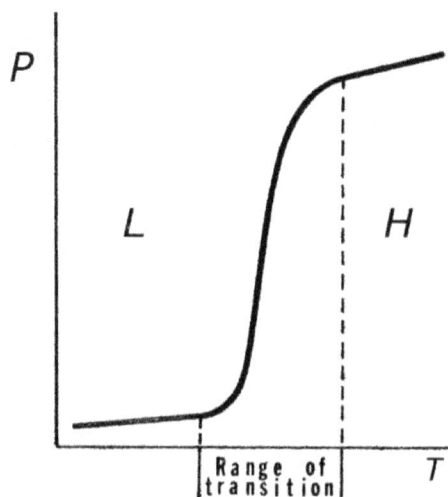

**Figure 2.** Typical "sigmoid" plot of a physical property $P$ upon heating through a phase transition. Its delineation in the temperature range of transition is actually determined by the relative quantities of H- and L-phases in the sample due to non-simultaneous nucleation

## 7. Hysteresis Loop of a Phase Transition

The AE and GD curves in Fig. 3 represent mass fraction $m_H$ of the H-phase. The sigmoid curve $P(T)$ in Fig. 2 simply delineates the mass ratio of the phases in the two-phase range. In a proper scale it becomes the right part of the hysteresis loop (AE in Fig. 3). Existence of the right part necessarily means that the left part (GD) would be found in the reverse run. It is important to note that the phase transition can be reversed only after $T$ is lowered below $T_0$, and even further to exceed a certain threshold range of stability, to activate an L-nucleus. The hysteresis loop in a small single crystal particle is rectangular. The sigmoid shape of the plots AE and GD is for the systems of many particles. It is indicative of two factors acting in opposite directions as the temperature rises. They are: (1) increase in the number of suitable nucleation sites per unit mass, and (2) decrease of the mass of the original phase. The former factor dominates in the initial stage, and the latter in the final stage of the phase transition.

## 8. When Hysteresis is Small

There are circumstances when hysteresis in solid-state phase transitions (and any other structural rearrangements in solids for that matter) can be especially small. Two such cases will be outlined here, but they are presented in greater detail in[2 (Sec. 2.8 and 2.9.3.3)], as well as in[5,7,16].

One case is nucleation in layered crystals A layered structure has strongly bounded, energetically advantageous two-dimensional (2-D) units − molecular layers, while the interlayer interaction is weak. Change from one polymorph to the other mainly involves the mode of layer stacking. Real layered crystals always have numerous defects resulted from imprecise layer stacking. Most of these defects are minute microcavities in the form of wedge-like interlayer cracks

concentrated at the crystal faces. In such a microcavity there is always a point where the gap has the optimum width for nucleation. There the molecular relocation from one side of interface to the other occurs with no steric hindrance and, at the same time, with the aid of attraction from the opposite wall. In view of a close structural similarity of the layers in the two polymorphs, this nucleation is *epitaxial*. There is a simple answer to why the temperature hysteresis $\Delta T_n$ in *epitaxial* phase transitions is small. Due to the abundance of wedge-like microcracks, there is no shortage in the nucleation sites; at that, the presence of a substrate of almost identical surface structure acts like a "seed". As a result, only small overheating or overcooling is required to initiate phase transition. Without a scrupulous verification, these phase transitions may seem "instantaneous", "without a hysteresis", "cooperative", "displacive", "second-order", *etc.*, but they are still a nucleation-and-growth.

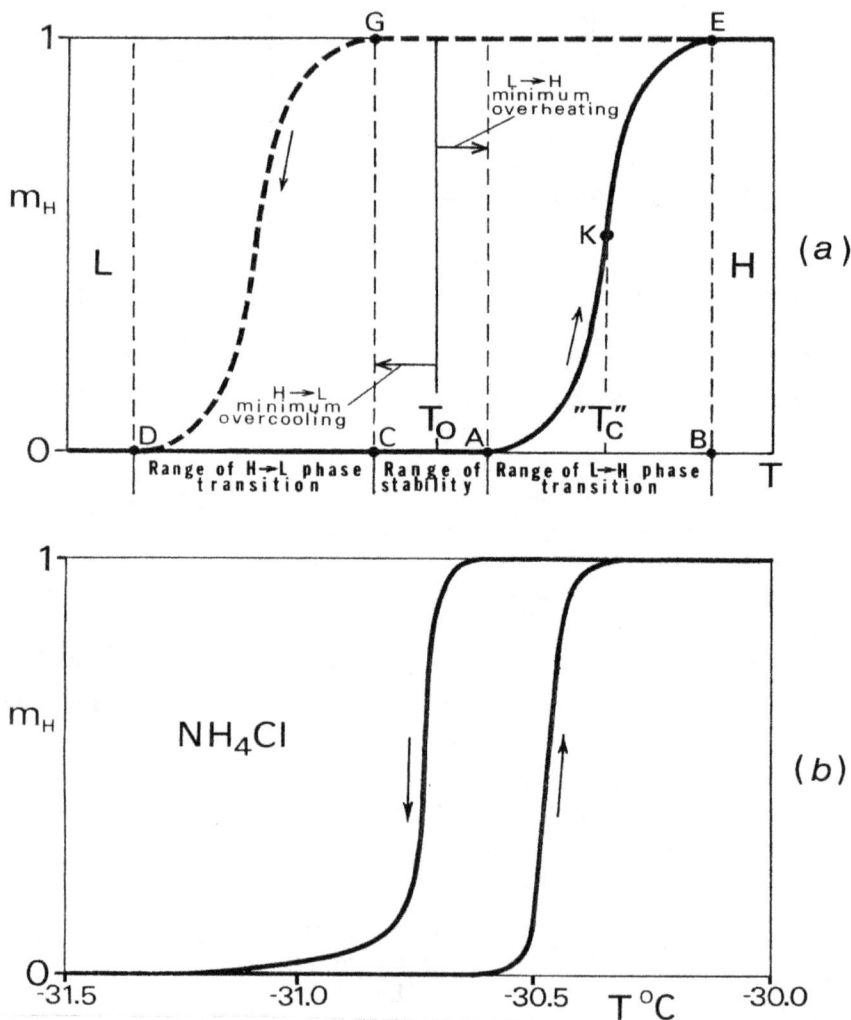

(a) (Schematic) "Sigmoid" curves AE and GD, each representing $m_H$ in the heterophase (L+H) temperature range of transition. Together they form a hysteresis loop DAEGD. Range of stability CA consists of two threshold lags too small to activate nucleation within. The inflection point K is not a "critical point" (or "Curie point"). It marks the temperature of the maximum number of activated nucleation sites;

(b) (Experimental) The hysteresis loop in $NH_4Cl$ (Dinichert[15])

**Figure 3.**    Hysteresis loop of a solid-state phase transition ($m_H$ is mass fraction of H phase)

Besides the hysteresis caused by formation of a 3-D nucleus to initiate the structural rearrangement, there is another type of hysteresis, much smaller, related to propagation of the immerged interfaces. It has been shown that its advancement in the normal direction also requires some overheating/overcooling. This hysteresis involves a 2-D nucleation. The 2-D nuclei form heterogeneously as well, but there is a significant difference in their function. While only a single 3-D nucleus at an *optimum microcavity* is needed to start a transition, a sufficient concentration of appropriate defects is required to keep the interface moving. These defects are also microcavities, but smaller, although not just individual vacancies. They were named *vacancy aggregations*. One such 2-D defect acts as a nucleus in reformation of a single molecular layer of the crystal and disappears when it is completed. Then another 2-D defect is needed to build next layer.

Existence of two kinds of nucleation − and the associated two-level hysteresis − means that a phase transition, ones started by a 3-D nucleus, may continue at a lower overheating/overcooling, provided the temperature is kept outside the 2-D nucleation threshold.

Another consequence relates to the exact value of $T_0$. It is not enough to find it as the temperature when the interface does not move in any direction. The two phases are not in dynamic equilibrium at their interface. The $T_0$ can be determined only approximately as being within the temperature range consisting of the positive and negative 2-D nucleation threshold lags.

# 9. Magnetization and Its Hysteresis

In the following we will use term *magnetization* towards any change in the magnetic state of a material. We start with two statements.

(1) Definition: "A phase, in the solid state, is characterized by its structure. A solid-state *phase transition* is therefore a transition involving a change of structure, which can be specified in geometrical terms" (Megaw,[17]).

(2) Orientation of a spin in the magnetic material is uniquely bound to the orientation of its atomic carrier[2 (Chapter 4)] and[10,18].

Let us assume a phase diagram where the free energy $F$ of a ferromagnetic material is presented in coordinates of the control parameters $T$ (temperature), $H$ (applied magnetic field), and others. Application of any one will move $F$ over the diagram and can cause it to intersect a border between the areas of the phase stability, giving rise to a phase transition − a *structural* phase transition in accordance with the Statement 1,. The molecular mechanism of the phase transition does not depend of which control parameter affected $F$ value; the mechanism will be the same, namely, a change of the crystal structure by nucleation and growth. That will bring about a change in the spatial positions and orientations of the spin carriers and their spins as well. In other words, ferromagnetic phase transition is *accompanied by magnetization*. The hysteresis of that structural rearrangement, caused by nucleation, will be *accompanied by magnetization hysteresis*.

Magnetization not necessarily related to phase transitions: it takes place every time magnetic field is applied to a ferromagnetic. There is a certain difference in how the control parameters $T$ and $H$ act on a ferromagnetic crystal. Temperature $T$ is a scalar. It is responsible for the level of thermal vibrations of all constituting particles and thus affecting their mutual bonding. The applied field $H$, on the other hand, is a vector acting only on the spin carriers over the direction of their spins (see Statement 2). Considering that magnetic interaction constitutes only small part of the crystal free energy[2 (Chapter 4)] and[18], it can be said qualitatively that $H$ is a relatively weak control parameter. If the directions of $H$ and spins differ, a readjustment can occur to lower $F$ by making spins coincide with, or be closer to, the $H$ direction. This is a *magnetization* not involving phase transition. Finding how it materializes is a high priority issue *per se* and the key to explanation of magnetic hysteresis.

The conventional accounts for the magnetization is vague[19]. It is usually described, without going into details, as a "domain rotation". No attempts are made to theoretically justify the experimental fact that magnetization is realized, at least in most cases, by moving interfaces. The second way of magnetization − by spin rotation in the original crystal structure remaining intact − is believed to also exist.

The new fundamentals of ferromagnetism[2,18] account for the magnetization and its hysteresis in a straightforward and coherent way by simply accepting that orientation of a spin is bound to its atomic carrier (Statement 2 above). It follows immediately that *magnetization* (spin turning) can be realized only by turning the spin carriers. The only way it can be accomplished in a crystal is its complete reconstruction by nucleation and interface propagation. While it is not a phase transition, it is realized like phase transitions do - by nucleation and growth, revealing the generality of nature processes.

Conclusions: (a) Magnetization by applied magnetic field is realized by the nucleation-and-growth structural rearrangement, (b) Any spin reorientation results from a reconstruction of the whole crystal itself, producing the new crystal structure identical to the initial, but in the new spatial orientation, (c) The alternative magnetization mechanism, assuming spin rotation *in* the crystal structure, cannot exist, and (d) *The answer to "What causes magnetic hysteresis?" is: the nucleation lags of the underlying structural rearrangement.*

Now we will limit the term "magnetization" to only designate the process not involving phase transition. It is accepted as an experimental fact that motion of domain boundaries is the main mechanism of magnetization, and that the hysteresis is lags in that motion and lags in formation of new domains. But the question why magnetization is localized on the domain boundaries has not been raised. The cause of the lags has not been identified. Besides, it is erroneously accepted that magnetization can also occur by "rotation" without motion of the domain boundaries. A possible relationship between the magnetic hysteresis loops $M = f(H)$ and the "structural" ones $P = f(T)$ in solid-state phase transitions (Fig. 2 and 3) has not attracted due attention. These failures were rooted in the interpretation of the lags as those of a *magnetic rearrangement in the crystal structure*, rather than a *rearrangement of the crystal structure* itself.

## 10. Rectangular Hysteresis Loop of Magnetization

Hysteresis loops of magnetization $M$ in external alternating magnetic fields $H$ are a prominent feature of ferromagnetic materials. When analyzing them, one has to take into account whether

- the sample is a single (single-domain) crystal, polydomain crystal or polycrystal,
- the magnetic field $H$ is applied in the "easy" or any other direction,
- the $H$ strength is sufficient to magnetize the sample to saturation,
- the loop is quasi-stationary, or not because it was recorded in a high frequency alternating field.

The actual shape of the hysteresis loops varies depending on these conditions, but, like their phase transition counterparts, they can be entirely accounted for in terms of the structural categories of nucleation and growth.

In the classical experiments by Sixtus and Tonks (S & T), described in a number of sources ([e. g.,[9,19]), the experimental arrangement allowed to investigate a magnetic hysteresis loop free of complicating side effects. There the sample - ferromagnetic wire - was turned to a single crystal magnetized to saturation $M_S$, while the magnetostriction effects and internal strains were eliminated. As a result, the sample exhibited a rectangular hysteresis loop as in Fig. 4. Reducing $H'$ to zero and even applying negative field $H'' < H_n$, where $H_n$ is the magnetic field necessary to start remagnetization, does not affect the $M = M_S$ value. The line A-B is horizontal. An additional applied negative field $\Delta H$, so that $|H'' + \Delta H| > H_n$ would trigger formation of a nucleus with $180°$-reversed magnetization, followed by the fast propagation of a domain interface over the whole sample. The line BC is vertical. Speed of the interface propagation was measured by different authors; it varied depending on the sample and on the strength of the magnetic field. The maximum speed was well below of what can be expected from a "magnetization wave". The magnetic field $H_n$ needed to create the nucleus was called "starting field" ("nucleation field" would be a better name). A somewhat weaker field, called "critical field", was sufficient to keep the interface moving.

Setting aside the shortcomings of the interpretation of the S & T experiments in[18], we will look at the subject in terms of the structural nucleation-growth concept. Remagnetization is not just a "wave of magnetization reversal": change in the $M_S$ direction occurs by the rearrangement of crystal structure at the interface. One may

argue that the crystal structures on sides of the domain interface are the same and, by definition, are not different phases. This argument is valid when the variable affecting the crystal free energy is a scalar, such as temperature or pressure, but magnetic field H is a vector. The free energies of two structurally identical domains differently oriented in the magnetic field are not the same, which is the driving force of structural rearrangement at the domain interfaces.

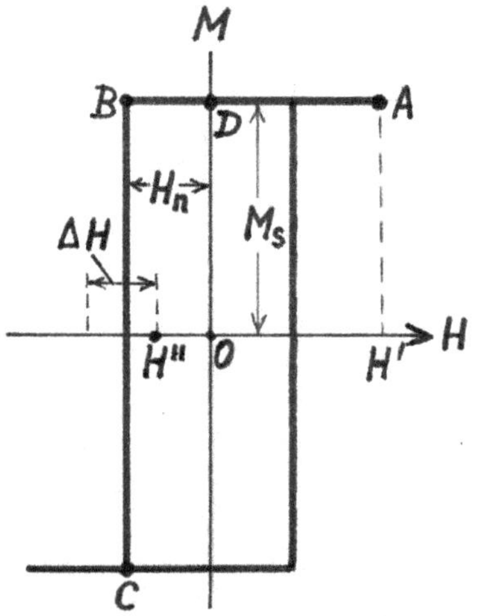

**Figure 4.** Ferromagnetic rectangular hysteresis loop. See text for symbols and description

The next point to clarify is the reason why $M_S$, achieved at the strongest positive magnetic field $H'$, remains unchanged after $H'$ is reduced to zero and even farther into the negative side (horizontal line A-D-B in Fig.4). The crystal structure is stable over the region A-D, indeed. In the region D-B, on the contrary, the sample is in the unstable state, since the direction of its magnetization is opposite to $H$. It remains quasi-stable simply because no structural change can occur without nucleation, i.e., until the negative field is sufficiently strong to increase the instability to the point when a structural nucleus of the opposite magnetization appears. This "starting field" $H_n$ has the same function as the overheating / overcooling in initiating temperature phase transitions. Considering that the $H_n$ value is "pre-coded" in a structural defect, it is not exactly reproducible in different samples. This behavior is no different from the temperature solid-state phase transitions: an energetically unstable phase remains quasi-stable until conditions for the formation of a 3-D nucleus are provided.

Since the domain interface motion was regarded in literature a "wave of magnetization reversal", there was a problem to explain why speed of this wave is too low. What really takes place at the domain interfaces is not a "wave", but a structural rearrangement. Structural phase transitions provide answer to the questions of why some excessive magnetic field ("critical field"), lower than a "starting field", is still required to keep the domain interface moving. The molecular mechanism of structural rearrangement at the domain interfaces is the same as in the structural phase transitions described above. It involves 3-D nucleation to start and 2-D nucleation to continue the process of magnetization. More detailed description of the domain boundaries and the structural rearrangement at them can be found in[2 (Sec. 4.9]. The nucleation lags of that structural rearrangement are the cause of the ferromagnetic and ferroelectric hysteresis loops.

# 11. Typical Hysteresis Loop of Magnetization

The rectangular hysteresis loop (Fig. 4) is at the basis of all ferromagnetic hysteresis loops. Only nucleation and growth are involved in its formation. The conditions for the rectangular loop to form are: a single-domain crystal, elimination of the magnetostriction adverse effect, a sufficiently strong magnetic field applied parallel/ antiparallel to the direction of spontaneous magnetization, quasi-stationary recording. The shape of a typical quasi-stationary ferromagnetic hysteresis loops, like in Fig. 5 (only its upper part is shown), always deviates from being rectangular to one or another degree. The overall cause for the "typical" loop to not be a rectangular is, evidently, that at least some of those conditions are not satisfied. As a rule, sufficient relevant information does not accompany real hysteresis loops, if at all. These loops are usually related to polycrystals, the fact being given little or no attention, much less properly taken into account.

Not infrequently the illustrative hysteresis loops look like the one in Fig. 6. They have such shape when being not quasi-stationary due to recording in fast alternating fields, instead of recording slowly or by point-by-point. If the applied field $H$ changes too fast, the domain interfaces do not have enough time to reach their quasi-stable positions corresponding to the $H$ amplitude. The shape of such loop depends on the frequency of the alternating field. Besides, the relaxation time of the internal strains caused by the magnetostriction is too short and is a function of the frequency as well.

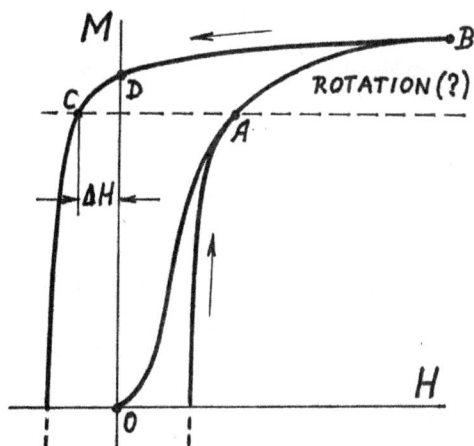

**Figure 5.** A typical remagnetization hysteresis loop (its lower part is omitted). See text for explanation of its particulars. Its part over the dotted line was erroneously claimed to be due to spin rotation in the structure

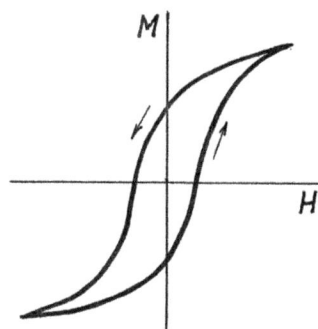

**Figure 6.** The type of a ferromagnetic hysteresis loop frequently used to illustrate the phenomenon. Such loops are not quasi-stationary and therefore do not fit for analyzing their shape

Only quasi-stationary loops are fit to be analyzed. Their detailed analysis[2, (Sec. 4.13.3)] goes beyond the scope of this article. In brief, their conventional interpretation (Bozorth[19]) was inadequate. A major issue is whether magnetization can occur by a "domain rotation". This magnetization process is claimed to take place in the part of loop marked "rotation (?)" in Fig. 5. According to Bozorth, at point A the magnetization stage owing to motion of the domain boundaries is completed; the magnetic moments of all domains in the sample became uniformly aligned (magnetically saturated) in the "easy" direction of the crystal; farther magnetization in the $H$ direction (from A to B) proceeds by a "reversible rotation" of the magnetic moments from the "easy" direction into the direction of the applied magnetic field $H$. It cannot be so, however. The polycrystalline material was treated as if it was a polydomain "single" crystal. Even more indicative is that the alleged rotation of $M_S$ from the "easy" direction by the magnetic field can only be elastic, because the crystal forces will try to return $M_S$ to the "easy" direction. In other words, magnetization by "rotation" has to be reversible, and Bozorth called it as such. But the actual process is not reversible, as evident from the fact that the B → C does not follow B → A. There are a number of reasonable causes, discussed in[2], to account for the ascending M from A to B. Impossibility of spin rotation in a crystal lattice is a major point of the new fundamentals of ferromagnetism and rooted in the fact that spin directions are fixed in their particles and therefore fixed in the crystal structure (see Section 9).

The magnetization B → D shows that a major portion of the structural rearrangements that occurred on the way from A to B is retained, but there is some regression causing the ensuing slope. Whichever processes led to the magnetization A → B, it was accompanied by accumulation of internal strains opposing this magnetization. A subsequent decrease in the $H$ strength allows strains to relax by means of structural readjustments at the expense of the magnetization. The strains can be eliminated by annealing the sample under the conditions marked by point B. In this way the curve B → D → C can be flattened, even made horizontal. The moderate $M$ decline over B → D → C is not a remagnetization yet. It begins only after $H$ changes its sign to the opposite and exceeds a certain threshold $-\Delta H$ to initiate nucleation of the oppositely oriented domains. But the sample is still polycrystalline and, contrary to a rectangular loop where a single nucleation act caused a propagation of the domain interface over the whole sample, this time one nucleation act affects only one crystal grain. The "starting fields" $H_n$ are different in different grains. The process would not proceed without $|H|$ increases. Still, this is the most effective magnetization phase $|dM/dH| = $ max, ending at the point equivalent to point $A$.

A common misconception should be dispelled regarding the role of crystal defects in a magnetization process and formation of hysteresis loops. The defects were always considered only as an obstacle to the motion of domain boundaries. In fact, their role is twofold. In a defect-free crystal neither motion, nor even formation of the boundary is possible. We can imagine a ferromagnet exhibiting a very high coercive force because its crystal structure is "too perfect". Indeed, the shortage of adequate defects for nucleation in very small ferromagnetic particles requires very strong fields for their remagnetization. On the other hand, different kinds of crystal defects that are not suitable to serve as heterogeneous nucleation sites may hamper propagation of domain boundaries.

## 12. Ferroelectric Hysteresis and Hysteresis Loops

Almost the entire description of ferromagnetic hysteresis and hysteresis loops is directly applicable to the ferroelectric hysteresis and its loops of repolarization in electric fields $E$. Only two points should be noted.

The *orientational* polarization (*i. e.*, by "rotation") is not possible as well. The difference is that here it is self-evident. The orientation of the electric dipoles in polar dielectrics is an element of their crystal structure. Reorientation of the dipoles in a ferroelectric crystal can occur only by rearrangements of crystal structure at the domain interfaces. Ferroelectrics can be polarized to saturation $P_S$ only in the direction determined by the crystal structure, and not in the arbitrarily chosen direction of the applied field $E$. Achieving $P_S$ in polycrystalline ferroelectrics in any $E$ direction is conceivable only through growth of the grains that happened to be polarized in the $E$ direction.

Another feature is the *induced polarization*. While spontaneous magnetization $M_S$ by itself does not appreciably depend on $H$, a spontaneous polarization $P_S$ depends on $E$ to some extent. It is due to the fact that the two electric charges of a ferroelectric dipole are spatially separated in the crystal unit cell. The induced polarization adds to the polarization caused by the structural domain rearrangements and is noticeably present in the hysteresis loops. Specifically, the saturation polarization $P_S(E)$ continues to grow even after all the dipoles are parallel. The induced polarization is strictly reversible and has therefore nothing to do with hysteresis, even though it somewhat affects the hysteresis loop shape.

For example, a ferroelectric hysteresis loop cannot be quite rectangular.

## 13. Conclusions

The cause of hysteresis in solid-state reactions, not previously identified, was experimentally established. The key method of the investigation was optical microscopy of phase transitions in good transparent single crystals. The exclusive cause of the hysteresis is found to be *nucleation lags* due to the features of the nucleation quite different from the classical "random fluctuation" model. The nuclei are located in specific crystal defects - microcavities whose size and shape determine the lags values. The hysteresis in magnetic and ferroelectric materials results from the underlying structural changes. The acquired information on the hysteresis cause and features was applied to interpretation of structural and magnetic hysteresis loops.

## REFERENCES

[1] L.D. Landau and E.M. Lifshitz, *Statistical Physics*, Addison-Wesley (1969).

[2] Y. Mnyukh, *Fundamentals of Solid-State Phase Transitions, Ferromagnetism and Ferroelectricity*, Authorhouse, 2001[or 2nd (2010) Edition].

[3] Y. Mnyukh, Second-order phase transitions, L. Landau and his successors, Am. J. Cond. Mat. Phys. 2013, 3(2): 25-30.

[4] Y. Mnyukh, N.A. Panfilova, Polymorphic transitions in molecular crystals - 2. Mechanism of molecular rearrangement at 'contact' interface, J. Phys. Chem. Solids 34 (1973) 159-170.

[5] Y. Mnyukh, Mechanism and kinetics of phase transitions and other reactions in solids, Am. J. Cond. Mat. Phys. 2013, 3(4): 89-103.

[6] *The Science of Hysteresis: 3-volume set,* Ed. G. Bertotti and I Meyergoyz, Academic Press, 2006.

[7] Y. Mnyukh, Molecular mechanism of polymorphic transitions, Mol. Cryst. Liq. Cryst. 52 (1979) 163-200.

[8] M. S. Pierce, C. R. Buechler, L. B. Sorensen, S. D. Kevan, E. A. Jagla, J. M. Deutsch, T. Mai, O. Narayan, J. E. Davies, K. Liu, G. T. Zimanyi, H. G. Katzgraber, O. Hellwig, E. E. Fullerton, P. Fischer, and J. B. Kortright, Phys. Rev. B 75, 144406 (2007).

[9] S. V. Vonsovskii, Magnetism, vol. 2, Wiley (1974).

[10] Y. Mnyukh, The physics of magnetization, arxiv.org /abs/1101.1249.

[11] Y. Mnyukh, Ferromagnetic state and phase transitions, Am. J. Cond. Mat. Phys. 2012, 2(5): 109-115.

[12] A. I. Kitaigorodskii, Y. Mnyukh, Y. Asadov, 1965, Relationships for single crystal growth during polymorphic transformation, J. Phys. Chem. Solids, 26, 463-472.

[13] Y. Mnyukh, 1976, Polymorphic transitions in crystals: nucleation, J. Crystal Growth, 32, 371-377.

[14] K.C. Russell, Nucleation in Solids, in *Nucleation III*, Ed. A.C. Zettlemoyer, M. Dekker (1977).

[15] P. Dinichert, Helv. Phys. Acta 15 (1942) 462-475.

[16] Y. Mnyukh, N. A. Panfilova, N. N. Petropavlov, N. S. Uchvatova, 1975, Polymorphic transitions in molecular crystals - 3, J. Phys. Chem. Solids, 36, 127-144.

[17] H.D. Megaw, Crystal Structures: A Working Approach, Saunders Co. (1973).

[18] Y. Mnyukh, Ferromagnetic state and phase transitions, Am. J. Cond. Mat. Phys., 2012, 2(5): 109-115.

[19] R.M. Bozorth, *Ferromagnetism*, Van Nostrand (1951).

# Phase Transitions in Layered Crystals

It is demonstrated by analyzing real examples that phase transitions in layered crystals occur like all other solid-state phase transitions by nucleation and crystal growth, but have a specific morphology. There the nucleation is *epitaxial*, resulting in the rigorous orientation relationship between the polymorphs, such that the directions of molecular layers are preserved. The detailed molecular mechanism of these phase transitions and formation of the laminar domain structures are described and related to the nature of ferroelectrics.

# Phase transitions in layered crystals

Yuri Mnyukh

*76 Peggy Lane, Farmington, CT, USA, e-mail: yuri@mnyukh.com*

(Dated: May 21, 2011)

It is demonstrated by analyzing real examples that phase transitions in layered crystals occur like all other solid-state phase transitions by nucleation and crystal growth, but have a specific morphology. There the nucleation is *epitaxial*, resulting in the rigorous orientation relationship between the polymorphs, such that the directions of molecular layers are preserved. The detailed molecular mechanism of these phase transitions and formation of the laminar domain structures are described and related to the nature of ferroelectrics.

## 1. Layered structures

Specifics of phase transitions in layered crystals will be demonstrated by analyzing two examples: *hexamethyl benzene* (HMB) and *DL-norleucine* (DL-N). They differ in their properties, molecular shape, and crystal structure.

HMB, $C_6(CH_3)_6$, is an aromatic substance with flat circular "coin-like" molecules (Fig. 1a). This molecular shape allowed a very energetically advantageous close packing into pseudo-hexagonal molecular layers, a molecular plane coinciding with the layer plane [1,2]. All intermolecular bonding, both in the layers and between the layers, is of a Van der Waals' type. The minimum distance between the carbon atoms of the benzene rings in adjacent (*001*) molecular layers is larger than the sum 3.40Å of carbon Van der Waals' radii due to repulsion of the $CH_3$ groups. As a result, the interaction between the layers is weakened, giving rise to the layered structure (Fig. 1b).

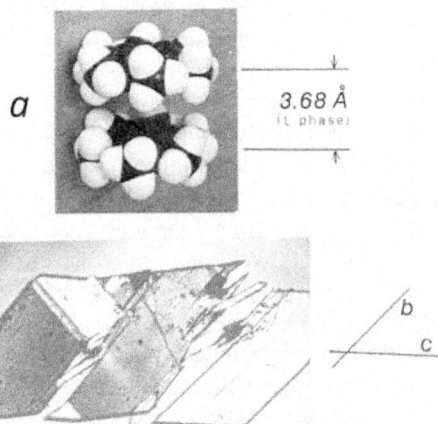

Fig. 1. Layered structure of hexamethyl benzene (HMB) crystals.
(a) HMB molecules in adjacent (*001*) layers, illustrating why HMB has layered structure: the molecular shape prevents sufficiently close interlayer packing.
(b) Lamination of a crystal along (*001*) when it is pricked with a needle.

DL-N is a short-chain aliphatic substance, $CH_3 \cdot (CH_2)_3 \cdot CHNH_2 \cdot COOH$, with a layered crystal structure typical of chain molecules, where the molecular axes are quite or almost perpendicular to the layer plane [3]. Each layer is bimolecular: the CNCOO groups of the molecules are pointed toward the center of the layer where they form a network of hydrogen bonds N-H... O [4] (Fig. 2). This central "skeleton" turns the bimolecular layer into a rather firm structural unit. The interlayer interaction is comparatively weaker, because it is of a purely Van der Waals' type, so the layer stacking is governed exclusively by the principle of close packing. As a result, both DL-N polymorphs have a pronounced layered structure of almost the same layers in different stacking..

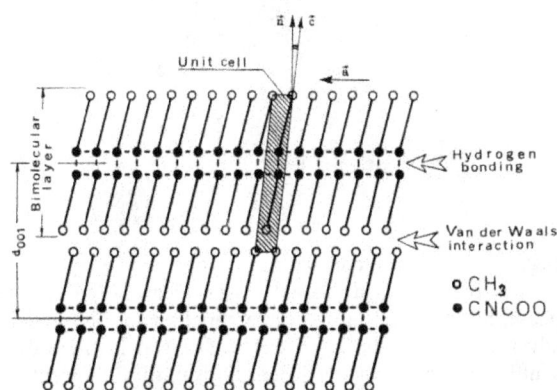

Fig.2. Characteristic features of the DL-norleucine (DL-N) crystal structure [4]. The layer spacing in the lower-temperature phase is $d_{001} = 16.03$ Å.

In general, a layered structure has strongly bounded, energetically advantageous two-dimensional units – molecular layers. Since the layer stacking contributes relatively little to the total lattice energy, the difference in the total free energies of the structural variants is small. This is a prerequisite for the polymorphism in layered crystals. Change from one polymorph to the other is mainly reduced to the mode of layer stacking.

The layers themselves only slightly modified under the influence of different layer stacking.

## 2. Nucleation-and-growth phase transitions

Prior to dealing with phase transitions in layered

Fig.3. Phase transition from low-to-high temperature phase (L→H) in *p*-dichlorobenzene (PDB). Four separate H single crystals of different orientations are growing within the L single crystal (background). Two largest ones have grown into one another as a result of competing for the building material that the surrounding L crystal is. Absence of orientation relationship between H crystals, as well as between them and L, is obvious. Evidently, the H *nuclei* had been oriented arbitrarily in the L lattice.

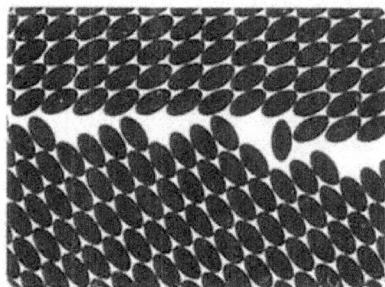

Fig. 4. Molecular model of phase transition in a crystal. The *contact* interface is a rational crystal plane in the resultant phase, but not necessarily un the initial phase. The interface advancement has the *edgewise* mechanism: it proceeds by shuttle-like strokes of small steps (kinks), filled molecule-by-molecule, and then layer-by-layer in this manner. (Crystal growth from liquids is realized by the same mechanism). Besides the direct contact of the two different structures, existence of the 0.5 molecular layer gap on average should be noted. It is wide enough to provide steric freedom for the molecular relocation (only at the kink), but it is narrow enough for the relocation to occur under attraction from the resultant crystal. More detailed description of the process and its advantages is given in Ref. 18 (Sec. 2.4.2 - 2.4.6 ).

crystals, the molecular mechanism of phase transitions.in non-layered crystals needs to be outlined. It has been revealed in the studies [5-17] summarized in [18]. It is a *crystal growth* involving nucleation and propagation of interfaces (Fig. 3), very much similar to

crystal growth from liquids. Molecular model of the interface and the molecule-by-molecule structural rearrangement leading to its propagation is shown in Fig. 4. The main feature of this mechanism is *a sufficient steric freedom of the molecular rearrangement still in the gravitation field of the new phase* [12].

The same principle is applied to the nucleation. It is not the classical fluctuation-based nucleation described in textbooks. It is pre-determined. The nuclei are located at specific crystal defects - microcavities of a certain optimum size. The microcavities provide sufficient steric freedom for the molecular relocation and, at the same time, assistance to that relocation by molecular attraction from the opposite side of the cavities. The distinctions in the size and shape of the microcavities determine both the individual nucleation temperature $T_n$ of each nucleation site and the crystal orientation of the new phase growing from it.

## 3. Phase transition in HMB

It had been reported [19] that HMB phase transition L→H at about 110 °C occurs "instantaneously" without changing the direction of light extinction when it was observed under crossed polarizers. The more detailed investigation of the transition with crystals of good quality [14,18] revealed the following. First, a more precise temperature $T_o$ (110.8 °C) corresponding to equal free energies of the polymorphs was established. Then it was found that the transition occurs not instantaneously, but by nucleation and gradual growth of the H phase. The nucleation occurred at crystal defects and exhibited hysteresis. The lags $\Delta T_{tr} = T_{tr} - T_o$ were greater in more perfect crystals. Nucleation could be initiated at any point by a touch with a needle. The interfaces were observable (Fig. 5). Their movement could be halted by lowering $\Delta T_{tr}$ to zero. Even small increases in $\Delta T_{tr}$ sharply accelerated interface motion (Fig. 6). At $\Delta T_{tr} > 2.5$ °C the interfaces advanced at the rate > 2 mm/sec. With nucleation lags of that order or higher (a realistic assumption), the phase transition in single crystals or grains of 0.4 mm size would be completed within 0.2 sec and appear instantaneous. Such an "instantaneous" transition is still $10^5$ -$10^6$ times slower than the velocity of elastic wave in a solid medium. The first transition was initiated at several points on the edges, continued by formation of thin H strips parallel to the cleavage, and then proceeded by a gradual width increase of the H bands denoted by shading between the arrows. The frontal advancement of the interfaces visible on the photographs was not, however, truly gradual; it rather proceeded by lateral strokes of very small steps along the interface lines (*edgewise* mechanism, like in Fig. 4), as a closer visual examination revealed.

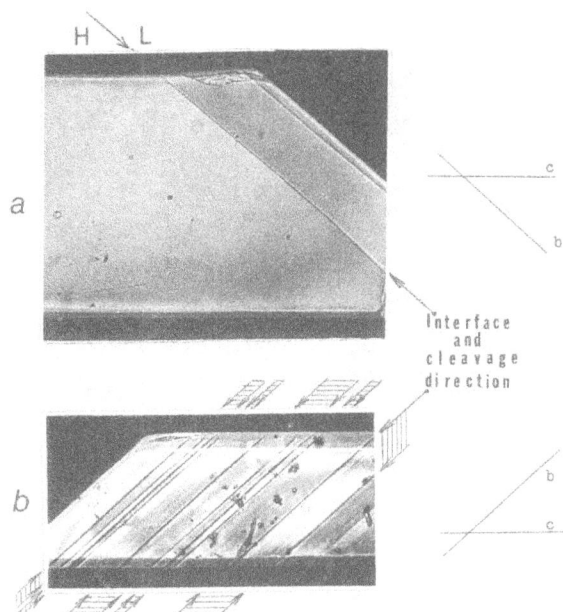

Fig.5. The interfaces (arrows) during phase transitions in HMB crystals. Crossed polarizers. Upon rotation of the microscope stage both phases were extinguished simultaneously.
(a) The first transition (L → H) in a fairly perfect single crystal. The interface remained parallel to itself and the cleavage planes.
(b)The last of the cyclic phase transitions
(L → H → L → H → ) L → H.

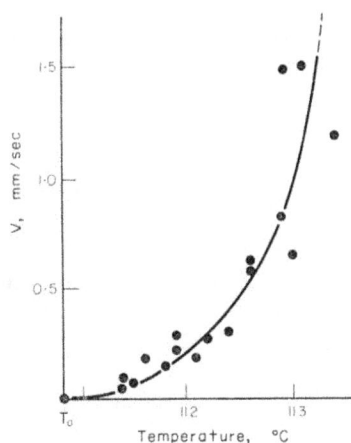

Fig. 6. The temperature dependence of velocity V of interface motion in HMB phase transition. $T_o = 110.8$ °C.

There can be no doubt in the nucleation-and-growth mechanism of the HMB phase transition. But its morphology (Fig. 5) differs from that in Fig. 3. This time the interfaces have always the same direction parallel to the molecular layers and cleavage.

Nucleation occurs at the layer edges and initially gives rise to formation of narrow wedges of the H phase with numerous growth steps along a wedge generating line (Fig. 7a). The wedges then penetrate through the crystal to form bands of the H phase seen in Fig. 5b. An important feature of this growth morphology is the *edgewise* mechanism as illustrated in Fig. 4. The molecular layers do not slip as a whole over one another.

Rather, *every layer is subjected to a complete rearrangement* (Fig. 7b). The x-ray Laue patterns confirmed a strict orientation relationship of the phases, while the optical examination with crossed polarizers made it clear that the layer orientation has not changed (Fig. 7c).

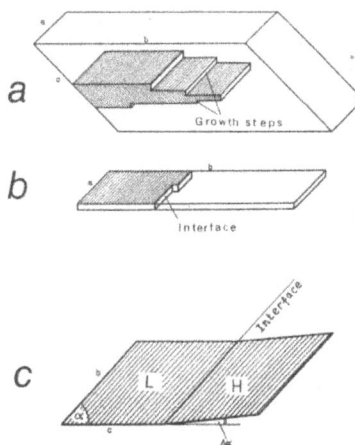

Fig. 7. Morphology of the HMB phase transition (schematic). See text.

## 4. Phase transition in DL-N

After discovery of the DL-N phase transition at ~117 °C [14] the first impression was that it may occur without hysteresis. In such a case it would not represent nucleation and growth. The task was to verify whether at least one of the sufficient indicators of the nucleation-and-growth mechanism is present. These indicators are [20] hysteresis, phase coexistence, interface motion, ability to initiate the transition by mechanical disturbance (note: in fact, these are different forms of one and the same indicator). To this end, smaller and more perfect single crystals were prepared. They were 0.5 to 2 mm size rhombus- or trapezium-shaped plates as thin as 0.02 to 0.1 mm. With these crystals and temperature control better than 0.1°C the nucleation hysteresis was detected, although it was rather small. Fig. 8 attests that all L → H transitions start at $T > T_o$, while H → L at $T < T_o$. Initially $|\Delta T|$ is about 0.8 °C, then decreases to stay at 0.2 °C level. In another experiment, $\Delta T_n$ was compared in pairs of one visually perfect and one less perfect single crystal of equal size. For each of the 10 pairs examined there was $|\Delta T_n|_{perf} > |\Delta T_n|_{imperf}$ by 0.3 to 0.8 °C. Finally, introduction of a mechanical defect initiated transition at a lower $\Delta T$. All the observations indicated nucleation at the crystal imperfections.

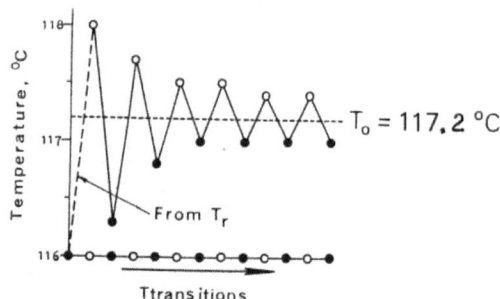

Fig. 8. Temperature hysteresis $\Delta T$ upon cyclic phase transitions in a DL-N crystal. The hysteresis is small and decreases with the number of the transitions to a low, but not zero, level.

When a (001) face was viewed in microscope under crossed polarizes, the direction of maximum extinction of the crystal was preserved after phase transition, thus showing a rigorous structural orientation relationship between the phases. Laue photographs taken before and after phase transition were almost indistinguishable, which proves both the rigorous orientation relationship and a strong similarity between the two structures. Then a set of the powder photographs as a function of temperature was taken and used to find the intralayer spacings $d_{200}$ and $d_{020}$ vs. temperature. There was a minute but noticeable (~1%) change in these spacings, which meant that the layers in the two phases are not completely identical.

The morphology of the phase transition was much like that in HMB. In an imperfect crystal, being actually a stack of weekly bound lamellae parallel to the molecular layers, the linear interfaces of one and the same direction moved separately and poorly coordinated in different lamellae. But in the relatively perfect crystals consisting of only few lamellae the linear interfaces were sharp. Their motion could be controlled by carefully regulated temperature. During this process every lamella was divided by the moving interface in two phases.

Finally, the long spacing $d_{001}$, which is a characteristic of the layer stacking, was observed on a screen and photographed upon heating over the transition range (Fig. 9). *Two distinct phases of different layer stacking coexisted throughout a range of transition.* The L phase was represented by the interlayer spacing $d_{001}=16.5$Å. Upon heating, the second line representing the interlayer spacing $d_{001} =17.2$ Å of the H phase

appeared. Its intensity gradually increased from zero to a maximum at the expense of the intensity of the L line until the latter was completely extinguished. This experiment decidedly refuted any idea of a gradual qualitative change.

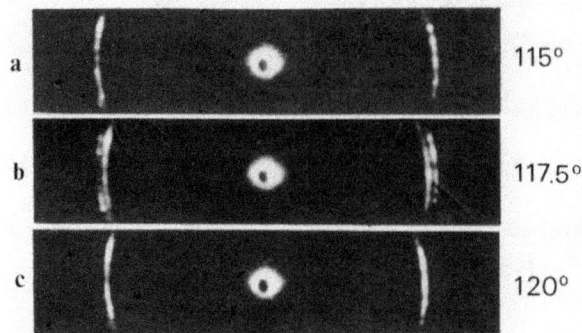

Fig. 9 Change in the interlayer spacing $d_{001}$ upon the L → H phase transition in a DL-N powder specimen. The photographs were taken from the screen of a device for direct viewing small-angle X-ray patterns.
(a) Before the transition. $d_{001} = 16.5$ Å.
(b) During the transition. Two separate lines coexist, each representing one of the phases. It was visually observed that the intensity of the H line was gradually increasing at the expense of the L line. Thus, the H *quantity* in the heterophase specimen was increasing, and L decreasing over the temperature range. This experiment visualizes how the apparent "continuous" and "displacive" phase transitions really proceed.
(c) After the transition; $d_{001}$ has increased by 4.1%.

## 5. Epitaxial nucleation in interlayer microcracks

HMB and DL-N differ in the molecular shape and chemistry, but exhibit similar features of their phase transitions not found in those described in Section 2. *This is due to their layered crystal structure.* These transitions occur by nucleation and growth as well. Nucleation requires presence of optimum-sized microcavities In layered structures the interlayer interaction is weak on definition. In practice, layered structures always have numerous defects of imprecise layer stacking. Most of these defects are minute wedge-like interlayer cracks on the crystal faces as viewed from the side of layer edges. In such a microcavity there always is a point where the gap has the optimum width for nucleation. There the molecular relocation from one wall to the other occurs with no steric hindrance and, at the same time, with the aid of attraction from the opposite wall. In view of the close structural similarity of the layers in the two polymorphs, *this nucleation is epitaxial.* In accordance with this description, the nucleation was indeed seen initiated at those crystal faces (Fig. 5 and 7). Fig. 10 is a schematic of initial stage of the epitaxial phase transition. Orienting effect by the substrate ensures preservation of the direction of molecular layers.

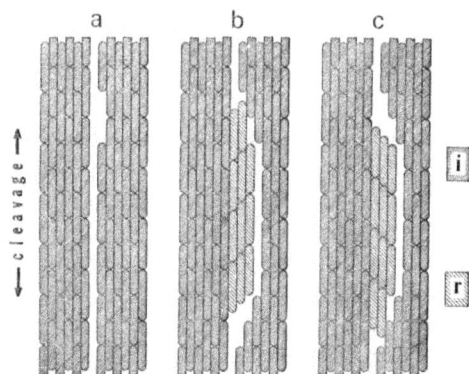

Fig. 10. Initial stage of epitaxial phase transition (schematic).
(a)  A lattice defect in the form of a submicroscopic crack parallel to the cleavage.
(b)  Oriented embryo of the resultant phase, formed by consecutive transfer of molecules from one side of the microcavity to the other.
(c)  Equally probable embryo in the "twin" orientation relative to its counterpart; it can form if the resultant lattice has a lower symmetry.

Finally, there is a simple answer to why the hysteresis $\Delta T_n$ in epitaxial phase transitions is small (Fig. 8) as compared with non-epitaxial transitions. Due to the abundance of wedge-like microcracks, there is no shortage in the nucleation sites of optimum size; at that, the presence of a substrate of almost identical surface structure acts like a crystallization "seed". Therefore only small overheating or overcooling is required in order to initiate phase transition. Without a scrupulous experimental verification, the phase transitions in question may be taken for "displacive", "instantaneous", "cooperative", "soft-mode", "second-order", etc.

## 6. Displacive phase transition by nucleation and growth

The idea of *displacive* phase transitions was put forward in 1950's as a cooperative displacement of atoms/molecules in the crystal lattice without breaking their bonding. At that, a rigorous structural orientation relationship was assumed without saying. The alternative was "reconstructive" phase transitions if such structural modification could not be imagined without breaking bonds; how the latter can occur remained unknown. The classification was not based on experimental investigations of the process. It was simply assumed from comparisons of the initial and final structures. The comparisons frequently resulted in "hybrid" cases with some bonds being broken; those cases were deemed "displacive" anyway. The adjective

"displacive" is now loosely applied to the cases where the structures of polymorphs seem to be "sufficiently similar".

In terms of a structural comparison of their polymorphs, the phase transitions in HMB and DL-N are the most suitable candidates to be *displacive* by displacing the molecular layers over one another to the new layer stacking. Indeed, a strict orientation relationship there is an experimental fact, and the bonding inside the layers remains unchanged. But, speaking figuratively, nature does not take advantage of making their phase transitions *displacive*. The transitions are a crystal growth on every account: nucleation, moving interfaces, phase coexistence in a temperature range, hysteresis, and discontinuous change (2.4% and 6%) in the specific volumes. Every molecular layer undergoes a reconstruction, molecule by molecule, to build up a new layer of almost the same structure, but now in a different layer stacking. Even the cementing action of hydrogen bonding inside the DL-N molecular layers does not prevent them from that reconstruction. The observational indicators of this kind of crystal growth is (a) rigorous orientation relationship of the phases, (b) a uniform direction of the interfaces, and (c) a relatively low hysteresis $\Delta T_n$. These features can be understood in terms of the *nucleation-and-growth* mechanism combined with the *nucleation epitaxy*. It is the most energy-efficient mechanism, considering that it needs energy to relocate only one molecule at a time, and not the myriads of molecules at a time as any cooperative change requires.

## 7. Laminar domain structures

It had been noticed [21-23] that phase transitions in layered crystals produce L phase in different equivalent orientations, approximately in equal quantities. The phenomenon was interpreted as the consequence of a *displacive* mechanism acting in different directions in different parts of the crystal. However, these phase transitions occur by *epitaxial crystal growth*. Formation of the laminar-domain structures is almost inevitable in epitaxial transitions if the emerging phase has a lower symmetry. The phenomenon is observed only in H $\rightarrow$ L transitions because it is L that has a lower symmetry. As illustrated in Fig. 10, an oriented L nucleus can appear with equal probability in two orientations related to one another as crystallographic twins. Fig. 11 shows this in more detail.

Growth of a nucleus gives rise to formation of a *laminar domain* of one or the other orientation. As a rule, phase transitions in layered crystals are multinuclear. Approximately one half of the laminar domains assumes the orientation No.1, and the rest assumes No.2. When the adjacent domains of the same orientation meet, they merge into a single domain. The laminar-domain structure with the strict alternation of No.1 and No.2 orientations, sketched in Fig.11b, emerges. Different domains are not uniform in thickness, but any two adjacent domains are related as crystallographic twins.

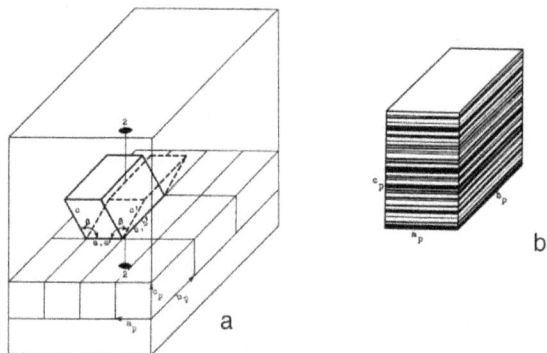

Fig. 11. Formation of "twin" domain structure in epitaxial phase transitions as a result of multinucleation of a lower-symmetry phase. The nuclei assume two equally probable orientations.

(a) Two equally probable orientations of a monoclinic lattice of the resultant phase in the initial crystal characterized by a rectangular lattice and cleavage (001). They can be brought into coincidence with a two-fold axis perpendicular to (001).

(b) Growth of each nucleus leads to formation of a lamina in one of the two possible orientations. The initial single crystal turns into a laminar structure of the domains of two alternating orientations shown as black and white.

## 8. Why ferroelectrics are not pyroelectrics

Pyroelectrics and ferroelectrics are both spontaneously polarized dielectrics, but only the latter have the ability to be polarized / repolarized by the applied electric fields. The difference can now be explained: only ferroelectrics have a layer structure. Nucleation in *ferroelectric↔paraelectric* phase transitions and in rearrangements of the laminar-domain systems is *epitaxial*. The *epitaxial* nucleation has a sufficiently low activation energy to be controlled by the applied electric field. The resultant structural change brings about a new state of polarization.

## References

[1]  L. Brockway and J.M. Robertson, *J. Chem. Soc.*, 1324 (1939).
[2]  T. Watanabe, Y. Saito and H. Chihara, *Sci. Papers Osaka Univ.*, No.2, 9 (1950).
[3]  Y. Mnyukh, *J. Phys. Chem. Solids*, **24** (1963) 631.
[4]  A. McL. Mathieson, *Acta Cryst.* **6,** 399 (1953).
[5]  A.I. Kitaigorodskii, Y. Mnyukh, Y. Asadov, *Soviet Physics - Doclady* **8** (1963) 127.
[6]  A.I. Kitaigorodskii, Y. Mnyukh, Y. Asadov, *J. Phys. Chem. Solids* **26** (1965) 463.
[7]  Y. Mnyukh, N.N. Petropavlov, A.I. Kitaigorodskii, *Soviet Physics - Doclady* **11** (1966) 4.
[8]  Y. Mnyukh, N.I. Musaev, A.I. Kitaigorodskii, *Soviet Physics - Doclady* **12** (1967) 409.
[9]  Y. Mnyukh, N.I. Musaev, *Soviet Physics - Doclady* **13** (1969) 630.
[10]  Y. Mnyukh, *Soviet Physics - Doclady* **16** (1972) 977.
[11]  Y. Mnyukh, N.N. Petropavlov, *J. Phys. Chem. Solids* **33** (1972) 2079.
[12]  Y. Mnyukh, N.A. Panfilova, *J. Phys. Chem. Solids* **34** (1973) 159.
[13]  Y. Mnyukh, N.A. Panfilova, *Soviet Physics - Doclady* **20** (1975) 344.
[14]  Y. Mnyukh *et al.*, *J. Phys. Chem. Solids* **36** (1975) 127.
[15]  Y. Mnyukh, *J. Crystal Growth* **32** (1976) 371.
[16]  Y. Mnyukh, *Mol. Cryst. Liq. Cryst.* **52** (1979) 163.
[17]  Y. Mnyukh, *Mol. Cryst. Liq. Cryst.* **52** (1979) 201.
[18]  Y. Mnyukh, *Fundamentals of Solid-State Phase Transitions, Ferromagnetism and Ferroelectricity*, Authorhouse, 2001 [or 2nd (2010) Edition].
[19]  V.L. Broude, *Zh. Eksp. Teor. Fiz.* **22**, 600 (1952).
[20]  Y. Mnyukh, http://arxiv.org/abs/1102.1085.
[21]  F.H. Herbstein, *Acta Cryst.* **18**, 997 (1965).
[22] I. Taguchi, *Bull. Chem. Soc. Jap.* **34**, 392 (1961).
[23] A.A. Boiko, *Sov. Phys.-Cryst.* **14**, 539 (1970).

# On Physics of Magnetization

Experimental observations reported in *Nature* are in accord with our concept that spin orientation is  fixed in the structure of the atomic spin carrier. It follows that magnetization (change of the spin orientation in a crystal matter) is realized by rearrangement of the crystal structure, and not by spin rotation in that structure, as the standard theory assumed. The rearrangement occurs by nucleation and propagation of interfaces according to the general *contact nucleation-and-growth* mechanism.

American Journal of Condensed Matter Physics 2016, 6(1): 17-19
DOI: 10.5923/j.ajcmp.20160601.03

# On Physics of Magnetization

Yuri Mnyukh

76 Peggy Lane, Farmington, CT, USA

**Abstract**　Experimental observations reported in *Nature* are in accord with our concept that spin orientation is fixed in the structure of the atomic spin carrier. It follows that magnetization (change of the spin orientation in a crystal matter) is realized by rearrangement of the crystal structure, and not by spin rotation in that structure, as the standard theory assumed. The rearrangement occurs by nucleation and propagation of interfaces according to the general *contact nucleation-and-growth* mechanism.

**Keywords**　Magnetization, Spin rotation, Nucleation-and-growth, Ferromagnetism

The experimental facts reported in *Nature* [1-3] are indicative of the need for changing the current interpretation of magnetization mechanism. Magnetization, whether it is caused by application of magnetic field or changing temperature, is presently believed to be a "rotation" ("switching", "reversal") of spins *in* the crystal which can remain intact. The new interpretation, initially put forward in 2001 [4], coherently accounts for all puzzling observations made in [1-3]. It states that *spin orientation is a permanent feature of the spin carrier* (atom, molecule), therefore magnetization inevitably involves turning the carriers. In other words, magnetization is inseparably linked with rearrangement of the crystal. Thus, in order to comprehend the real magnetization process, understanding of the molecular* mechanism of solid-state rearrangements is required.

The following is a synopsis of the general mechanism of structural rearrangements, deduced from the studies presented initially by the sequence of journal articles [5-18] and then summarized in the book [4].

- Rearrangements in a solid state are realized by crystal growth involving nucleation and propagation of interfaces. Neither ferromagnetic phase transitions (see below), nor ferroelectric phase transitions [19] are excluded from this rule. Not a single sufficiently well-documented example exists of this process being homogeneous (cooperative).
- The nuclei are located in specific crystal defects - microcavities of a certain optimum size. These defects contain information on the condition (*e.g.*, temperature) of their activation and the orientation of resultant crystal lattice. The nucleation can be epitaxial, in which case a certain orientation relationship between the initial and resultant structures is observed.
- The interface is a rational crystallographic plane of the resultant crystal lattice. It is named "contact interface" owing to direct molecular contact between the two lattices without any intermediate layer. The molecular rearrangement proceeds according to *edgewise* (or *stepwise*) mechanism (Fig.1) consisting of formation of "kinks" (steps) at the flat interface and filling them, molecule-by-molecule, until the layer is complete, and building successive layers in this manner.

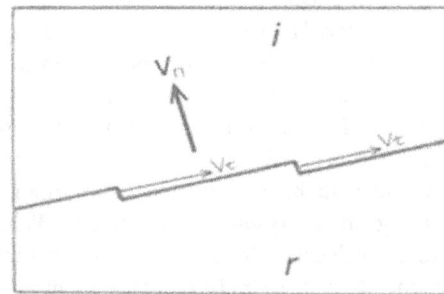

**Figure 1.**　The *edgewise* mechanism of phase transitions and any other rearrangements in solid state, such as at domain boundaries. The sketch illustrates the mode of advancement of interface in the **n** direction by shuttle-like strokes of small steps (kinks), filled by molecule-by-molecule, in the direction; *i* and *r* – are initial and resultant crystals, respectively. (A crystal growth from liquids is realized by the same manner). More detailed description of the mechanism and its advantages is given in [4]

Lavrov *et al.* [1] (LKA) have observed crystal rearrangement of a antiferromagnet by magnetic field. Three relevant aspects of the LKA work will be highlighted:

*First.* In terms of the conventional science the phenomenon itself must not exist. As LKA admitted, "the common perception [is that] magnetic field affects the orientation of spins, but has little impact on the crystal structure." But the structure did changed in their experiments. According to the new concept, however, structural

* Corresponding author:
yuri@mnyukh.com (Yuri Mnyukh)

Published online at http://journal.sapub.org/ajcmp

rearrangement is the only way of changing spin reorientation (i.e., of magnetization).

*Second.* LKA note that "one would least expect any structural change to be induced in antiferromagnet where spins are antiparallel and give no net moment". Nevertheless, such an unexpected phenomenon has took place. The reason for that seeming contradiction is rooted in the belief that spins in an antiferromagnet are strongly bound together by the Heisenberg's "exchange forces", therefore the external field H, which is weak in comparison, cannot deal separately with the parallel and antiparallel components of the spin system.

The legitimacy of the "exchange forces" theory was challenged in [4]. Even its initial verifications had to prevent its acceptance. The case in point is that the verifications have produced a *wrong sign* of the exchange forces. Despite this fatal defect, this theory was taken for granted. But Feynman [20] was skeptical at least, as seen from these statements: "When it was clear that quantum mechanics could supply a tremendous spin-oriented force - even if, apparently, of the wrong sign - it was suggested that ferromagnetism might have its origin in this same force", and "The most recent calculations of the energy between the two electron spins in iron still give the *wrong* sign", and "It is worse than that. Even if we *did* get the right sign, we would still have the question: why is a piece of lodestone in the ground magnetized?", and even "This physics of ours is a lot of fakery." The sign problem was carefully examined in a special review [21] and found fundamentally unavoidable in the Heisenberg model. It was suggested that the "neglect of the sign may hide important physics."

LKA actually dealt with the *antiferromagnet to ferromagnet structural phase transition.* There every second *spin carrier* was turned, so that its spin turned toward the direction of external magnetic field. Evidently, spins were strongly bound to their carriers rather than to each other.

*Third.* LKA observed a generation and motion of crystallographic twin boundaries and kinks moving along them, resulting in a crystal rearrangement. While their findings are inconsistent with the idea of spin "switching" or with any cooperative phase transition, they are in accord with the magnetization mechanism by crystal growth illustrated in Fig.1.

Novoselov *et al.* [2] recorded magnetization picture with a high resolution never before attained. They found that the ferromagnetic domain interface propagated by distinct jumps matching the lattice periodicity, the smallest being only a single lattice period. The "kinks" were running along the interface. The authors interpreted the interface movements as following the Peierls potential of crystal lattice and stated that further theoretical and experimental work is needed to understand the unexpected dynamics of domain walls. The phenomenon, however, had been described, predicted to be traced to the molecular level, explained and illustrated with a molecular model in [8, 11, 13, 17, 4] (see Fig.1 above). In fact, the same mode of interface propagation (running kinks and filling layer-by-layer) was observed by LKA [1] as well,

only on a more macroscopic level, and the fact that this led to a real crystal rearrangement was firmly established.

Tudosa *et al.* [3] estimated experimentally the ultimate speed of "magnetization switching" in tiny single-domain particles - an important issue in developing of magnetic memory devices. The speed turned out three orders of magnitude lower than was predicted and, besides, was not the same in the effected particles. The error of that prediction was hidden in the term "switching", in other words, in the assumption of a cooperative spin rotation *in* the current crystal structure. The lower speed had to be expected if magnetization is not a "switching", but occurs by nucleation and growth in every individual domain. Nucleation is heterogeneous, requires specific crystal defects and not simultaneous in different particles. It is nucleation that controls the magnetization of small single-domain particles [22].

The experimental observations presented in [1-3] provide strong evidence that magnetization is realized by structural rearrangement according to the specific *edgewise* mechanism involving nucleation and propagation of interfaces, rather than by spin rotation, switching or reversal in the same structure. The detailed description of the magnetization process without dipole rotation in the crystal structure is presented in [23].

# Appendix

In contemplating the problems of ferromagnetism and magnetization, it was least expected that this will lead to certain conclusions about the structure of a magnetic atom. They are: (a) The atom participates in the crystal structure as a particle of certain geometrical shape, and (b) The direction of its spin is permanently fixed in its structure.

Hopefully, these features will be noticed by the experts in atomic and nuclear physics.

* "Atomic" and "molecular" are interchangeable in the text.

---

# REFERENCES

[1]   A. N. Lavrov, Seiki Komiya and Yoichi Ando, Magnetic shape-memory effects in a crystal, Nature 418, (2002) 385.

[2]   K. S. Novoselov, A. K. Geim, S. V. Dubonos, E. W. Hill and I. V. Grigorieva, Subatomic movements of a domain wall in the Peierls potential, Nature 426, (2003) 812.

[3]   I. Tudosa, C. Stamm, A. B. Kashuba, F. King, H. C. Siegmann, J. Stöhr, G. Ju, B. Lu, The ultimate speed of magnetic switching in granular recording media, Nature, 428, (2004) 831.

[4]   Y. Mnyukh, Fundamentals of Solid-State Phase Transitions, Ferromagnetism and Ferroelectricity, Authorhouse, 2001 [or 2nd (2010) Edition].

[5]   Y. Mnyukh, Laws of phase transformations in a series of normal paraffins, J Phys Chem Solids, 24 (1963) 631.

[6]   A. I. Kitaigorodskii, Y. Mnyukh, Y. Asadov, The polymorphic single crystal – single crystal transformation in p-dichlorobenzene, Soviet Physics - Doclady 8 (1963) 127.

[7]   A. I. Kitaigorodskii, Y. Mnyukh, Y. Asadov, Relationships for single crystal growth during polymorphic transformation, J Phys Chem Solids 26 (1965) 463.

[8]   Y. Mnyukh, N. N. Petropavlov, A. I. Kitaigorodskii, Laminar growth of crystals during polymorphic transformation, Soviet Physics - Doclady 11 (1966) 4.

[9]   Y. Mnyukh, N. I. Musaev, A. I. Kitaigorodskii, Crystal growth in polymorphic transitions in glutaric acid and hexachloroethane, Soviet Physics - Doclady 12 (1967) 409.

[10]  Y. Mnyukh, N. I. Musaev, Mechanism of polymorphic transition from the crystalline to the rotational state Soviet Physics - Doclady 13 (1969) 630.

[11]  Y. Mnyukh, Molecular mechanism of polymorphic transitions, Soviet Physics - Doclady 16 (1972) 977.

[12]  Y. Mnyukh, N. N. Petropavlov, Polymorphic transitions in molecular crystals – I. Orientations of lattices and interfaces, J Phys Chem Solids 33 (1972) 2079.

[13]  Y. Mnyukh, N. A. Panfilova, Polymorphic transitions in molecular crystals – II. Mechanism of molecular rearrangement at 'contact' interface, J Phys Chem Solids 34 (1973) 159.

[14]  Y. Mnyukh, N. A. Panfilova, Nucleation in a single crystal, Soviet Physics - Doclady 20 (1975) 344.

[15]  Y. Mnyukh, N. A. Panfilova, N. N. Petropavlov, N. S. Uchvatova, 1975, Polymorphic transitions in molecular crystals - III. Transitions exhibiting unusual behavior, J. Phys. Chem. Solids, 36, 127.

[16]  Y. Mnyukh, Polymorphic transitions in crystals: nucleation, J Crystal Growth 32 (1976) 371.

[17]  Y. Mnyukh, Polymorphic transitions in crystals: nucleation Mol Cryst Liq Cryst 52 (1979) 163.

[18]  Y. Mnyukh, Polymorphic transitions in crystals: kinetics, Mol Cryst Liq Cryst 52 (1979) 201.

[19]  e.g., V. M. Ishchuk, V. L. Sobolev, Investigation of two-phase nucleation in paraelectric phase of ferroelectrics with ferroelectric–antiferroelectric–paraelectric triple point, J Appl Phys 92 (2002) 2086.

[20]  R. P. Feynman, R. B. Leighton, M. Sands, The Feynman Lectures on Physics v 2, Addison-Wesley New York, 1964.

[21]  J. H. Samson, 1995, Classical effective Hamiltonians, Wigner functions, and the sign problem, Phys Rev B 51, 223-233.

[22]  Sec.4 in Ref. 4: Nucleation in single-domain particles.

[23]  Y. Mnyukh, Magnetization of ferromagnets, Am J Cond Mat Phys 2014, 4(4): 78.

# Ferromagnetic State and Phase Transitions

Evidence is summarized attesting that the standard exchange field theory of ferromagnetism by Heisenberg has not been successful. It is replaced by the crystal field and a natural assumption that spin orientation is inexorably associated with the orientation of its carrier. It follows at once that ferromagnetic state is a property of the crystal structure and that both ferromagnetic phase transitions and magnetization must involve a structural rearrangement. The mechanism of structural rearrangements in solids is nucleation and interface propagation. The new approach accounts consistently for ferromagnetic state and its manifestations.

American Journal of Condensed Matter Physics 2012, 2(5): 109-115
DOI: 10.5923/j.ajcmp.20120205.01

# Ferromagnetic State and Phase Transitions

Yuri Mnyukh

Chemistry Department and Radiation and Solid State Laboratory, New York University, New York, NY 10003, USA

**Abstract** Evidence is summarized attesting that the standard exchange field theory of ferromagnetism by Heisenberg has not been successful. It is replaced by the crystal field and a natural assumption that spin orientation is inexorably associated with the orientation of its carrier. It follows at once that ferromagnetic state is a property of the crystal structure and that both ferromagnetic phase transitions and magnetization must involve a structural rearrangement. The mechanism of structural rearrangements in solids is nucleation and interface propagation. The new approach accounts consistently for ferromagnetic state and its manifestations.

**Keywords** Ferromagnetism, Heisenberg, Exchange, Sign Problem, Curie, λ-Anomaly, Phase Transition, First-Order, Magnetostructural, Nucleation, Interface, Hysteresis, Magnetization, Domain Structure, Barkhausen, Magnetostriction, Magnetocaloric, Ferroelectrics

## 1. Weiss' Molecular and Heisenberg's Electron Exchange Fields

Generally, ferromagnetics are spin-containing materials that are (or can be) magnetized and remain magnetized in the absence of magnetic field. This definition also includes ferrimagnetics, antiferromagnetics, and practically unlimited variety of magnetic structures. The classical Weiss / Heisenberg theory of ferromagnetism, taught in the universities and presented in many textbooks (e. g.,[1-4]), deals basically with a special case of the collinear (parallel and antiparallel) spin arrangement. The logic behind the theory in question is as follows. There is a spontaneously magnetized crystal (e. g., of Fe or Ni) due to a parallel alignment of the elementary magnetic dipoles. It remains stable up to its critical (Curie) temperature point when the thermal agitation suddenly destroys that alignment. It needed to be explained how the ferromagnetic state can be thermodynamically stable up to the really observed temperatures so high as 1042 K in Fe. It seemed unavoidable to suggest that the force holding the dipoles in parallel is the dipole interaction. Setting aside the probability that such interaction in Fe would rather cause mutual dipole repulsion than attraction, how strong must this interaction be? It followed from the Weiss' theory that it had to be about $10^4$ times stronger than the magnetic dipole interaction alone. The conclusion seemed undeniable: besides the magnetic dipole interaction, there is also interaction due to a much more powerful "molecular field" of unknown physical

nature. Heisenberg[5] accepted the Weiss' theory and developed its quantum-mechanical interpretation. His theory maintains that overlapping of the electron shells results in extremely strong *electron exchange interaction* responsible for collinear orientation of the magnetic moments. The main parameter in the quantum mechanical formula was *exchange integral*. Its positive sign led to a collinear ferromagnetism, and negative to a collinear antiferromagnetism. Since then it has become accepted that Heisenberg gave a quantum -mechanical explanation for Weiss' "molecular field": "Only quantum mechanics has brought about explanation of the true nature of ferromagnetism" (Tamm[2]). "Heisenberg has shown that the Weiss' theory of molecular field can get a simple and straightforward explanation in terms of quantum mechanics" (Seitz[1]).

## 2. Inconsistence with the Reality

General acceptance of the Heisenberg's theory of ferromagnetism remains unshakable to the present days.

Judging from the textbooks on physics, one may conclude that it is rather successful[6]. In these books and other concise presentations every effort was made to portray it as basically valid and a great achievement, while contradictions, blank areas, and vast disagreements with experiment are either omitted as "details" or only vaguely mentioned. As a result, a new student gets wrong impression about the real status of the theory. In general, the theory remains basically unchallenged. But the more detailed the source is, the more drawbacks are exposed. There are experts who pointed out to its essential shortcomings.

Bleaney & Bleaney[7]: "There is no doubt that ferromagnetism is due to the exchange forces first discovered by Heisenberg, but the quantitative theory of ferromagnetism contains many difficulties".

*Corresponding author:
yuri@mnyukh.com (Yuri Mnyukh)
Published online at http://journal.sapub.org/ajcmp

"We have a broad understanding of the outlines of ferromagnetic theory, but not of the details. The exchange interaction between two electrons cannot be calculated *a priori*. We cannot even be certain of its sign."

Belov[8]: ".Many important questions connected with the behavior of materials in the region [of ferromagnetic transition] remain unsettled or in dispute to the present time. These include ...the actual temperature behavior of the spontaneous magnetization near the Curie point, the causes of the 'smearing out' of the magnetic transition. the existence of 'residual' spontaneous magnetization above the Curie temperature, and the nature of the temperature dependence of elastic, electric, thermal, and other properties near the Curie point. It even remains unsettled what we should take to be the Curie temperature, and how to determine it".

"The theory of Weiss and Heisenberg cannot be applied to the quantitative description of phenomena in the neighborhood of the Curie point. Even for such a 'simple' ferromagnetic substance as nickel it is not possible to 'squeeze' the experimental results into the Weiss-Heisenberg theory".

Bozorth[3]: "The data for iron and for nickel [at low temperatures] show that the Weiss theory in either its original or modified form is quite inadequate".

"The Curie point is not always defined in accordance with the Weiss theory but in other more empirical ways..."

Crangle[9]: "It seems difficult to be convinced that direct exchange between localized electrons can be the main origin of the ferromagnetism in metals of the iron group".

Kittel[6]: "The Neel temperatures $T_N$ often vary considerably between samples, and in some cases there is large thermal hysteresis".

Feynman[10]: "Even the quantum theory deviates from the observed behavior at both high and low temperatures".

"The exact behavior near the Curie point has never been thoroughly figured out".

The theory of the sudden transition at the Curie point still needs to be completed."

"We still have the question: why is a piece of lodestone in the ground magnetized?"

"To the theoretical physicists, ferromagnetism presents a number of very interesting, unsolved, and beautiful challenges. One challenge is to understand why it exists at all".

The last statement is especially indicative, considering that it was the primary purpose of the Weiss' and Heisenberg's theories to explain why ferromagnetism exists at all. Moreover, it turned out that the exchange forces, as powerful as they assumed to be, do not physically participate in the actual ferromagnetic phenomena. Thus, Seitz[1] maintained that the "Heisenberg's model...is too simple to be used for quantitative investigation of the real ferromagnetic materials". Tamm[2] noted that "it is the usual magnetic interaction of atoms [rather than exchange interaction] that is responsible for such, for example, phenomena as magnetic anisotropy and magnetostriction". In this respect many other phenomena could also be mentioned: domain structure, magnetic hysteresis, magnetocaloric effect, Barkhausen effect, first-order magnetic phase transitions, magnetization kinetics, and more. Remarkably, the question why the exchange forces do not exhibit themselves in those phenomena has never been raised.

There are also other phenomena and facts the exchange interaction offers no reasonable explanation, if at all. Among them:

(A) The value of the exchange integral for Ni was found lower by about two orders of magnitude needed to account for its Curie temperature.

(B) A collinear order of the atomic magnetic moments in ferro, antiferro and ferrimagnetics represents only particular cases, while there is, in fact, a great variety of noncollinear magnetic structures as well. The exchange field was unable to provide a parallel alignment in those innumerable magnetic structures.

(C) There are materials where magnetic moments are too far apart to make any direct exchange possible. The appropriate electron shells in the ferromagnetic rareearth metals do not overlap. The 'exchange field' theory was expanded to those cases anyway, to become "superexchange".

(D) The actual speed of magnetization is well below of the theoretically expected.

(E) The exchange forces have the wrong sign.

# 3. The Sign Problem

Even the initial verifications of the Heisenberg's theory had to prevent its acceptance. The verifications have produced a *wrong sign* of the exchange forces. Feynman[10] was sceptical at least, as seen from these statements: "When it was clear that quantum mechanics could supply a tremendous spin-oriented force - even if, apparently, of the wrong sign - it was suggested that ferromagnetism might have its origin in this same force", and "The most recent calculations of the energy between the two electron spins in iron still give the *wrong* sign", and even "This physics of ours is a lot of fakery." The sign problem was later carefully examined in a special review[11] and found fundamentally unavoidable in the Heisenberg model. It was suggested that the "neglect of the sign may hide important physics.

# 4. Ferromagnetic Transitions Become "Magnetostructural"

In order to present a coherent picture of ferromagnetism, which is the purpose of this article, the molecular mechanism of ferromagnetic phase transition should be established. With this in mind, it will be helpful to trace the evolvement of views on ferromagnetic phase transitions. Initially it was everyone's belief that they are of the second order - a cooperative phenomenon with a fixed (Curie) temperature of phase transition. Kittel[6] used Ni as an example to state:

"This behavior classifies the usualferromagnetic ↔ paramagnetic transition as second order". In 1965 Belov wrote in his monograph "Magnetic Transitions"[8] that ferromagnetic and antiferromagnetic transitions are "concrete examples" of secondorder phase transitions.His work was devoted to the investigation of spontaneous magnetization and other properties in the vicinity of the Curie points.The problem was, however, how to extract these "points" from the experimental data which were always "smeared out" and had "tails" on the temperature scale, even in single crystals.

Vonsovskii[4] was still on that initial stage when stated that the theory of secondorder phase transitions provided an "impetus" to studies of magnetic phase transitions. But he already entered the second stage of the "evolvement" by recognizing that there are a number of the first-order ferromagnetic phase transitions. In his book about 25 such phase transitions were listed, still as rather "exotic". They were interpreted in the usual narrowformal manner as those exhibiting abrupt changes and/or hysteresis of the magnetization and other properties. Some of these firstorder ferromagnetic transitions Vonsovskii erroneously described as "apparent", where structural transitions occur before the ferromagnetic-to-paramagnetic transitions, but existence of genuine firstorder ferromagnetic transitions was also recognized. The puzzling fact of their existence led to the numerous theoretical and experimental studies surveyed in the book. The conventional theory was in a predicament: the Curie point was not a point any more, and was rather a range of points and, even worse, was a subject to temperature hysteresis. Attempts were made, with no success, to complicate the theory by making the exchange field dependent on the lattice deformation, interatomic parameters, energy of magnetic anisotropy, *etc*. The firstorder ferromagnetic phase transitions, so alien to the conventional theory, had to be accepted simply as an undeniable reality. *It was not realized that a firstorder phase transition meant nucleation and growth, and not a critical phenomenon.*

The number of recognized first-order ferromagnetic phase transitions continued growing. They were found to be of the fist order even in the basic ferromagnetics - Fe, Ni and Co[12-14]. This process was accompanied by the increasing realization of structural changes involved. A new term *"magnetostructural"* transitions has come into use to distinguish them from not being "structural". At the present time the quantitative ratio "magnetostructural/ second order" is dramatically shifting in favor of the "magnetostructural" phase transitions. The search with Google in June 8, 2011 produced

'second order ferromagnetic'....286,000 hits,
'first order ferromagnetic'.........926,000 hits,
'magnetostructural transition'...718,000 hits.

## 5. The Assumptions

The above trend is obvious, addressing us toward the conclusion that *all* ferromagnetic phase transitions are "structural", meaning they are always realized by nucleation and crystal rearrangements at the interfaces, rather than cooperatively. While this conclusion will formally remain our assumption, it is destined to be accepted as a fact. Designations of phase transitions as second order are always superficial. Not a single sufficiently documented example, ferromagnetic or otherwise, exists. This is because a nucleation-growth phase transition represents the most energy -efficient mechanism, considering that it needs energy to relocate only one molecule at a time, and not the myriads of molecules at a time as a cooperative process requires. Refer to[13,15].

The other assumption is: *the orientation of a spin is determined by the orientation of its atomic carrier.* Considering that the atomic carrier is an asymmetric entity, this simple assumption is more probable than ability of a spin to acquire different orientations in the same atom. These two assumptions represent the new fundamentals allowing to coherently account for ferromagnetic state and the numerous ferromagnetic phenomena. Knowledge of the actual molecular mechanism of nucleation-and-growth phase transitions will be necessary. Importantly, this will not require introduction of a "molecular field" of any kind in addition to the already existing chemical crystal bonding and magnetic dipole interaction.

## 6. Crucial Role of the Crystal Structure

Two opposing factors were considered by the Weiss' theory: the "molecular field" causing a parallel alignment of the ensemble of elementary magnets and the thermal agitation destroying this alignment. There the role of a crystal structure was implicitly reduced only to providing a positional, but not orientational, order to its magnetic dipoles. A system of atomic magnetic dipoles was a dipole system only. The objects of thermal agitation were the elementary magnets, and not the atoms carrying them. The *crystal field was overlooked.* There are powerful bonding forces combining molecules, ions, atoms, magnetic or not, into a crystal 3D longrange order, both positional and orientational. *It is the crystal field that imposes one or another magnetic order by packing spin carriers in accordance with the structural requirements.*

## 7. The Mechanism of Nucleation and Growth Phase Transitions

The following is a synopsis of the general mechanism of solid-state phase transitions and other structural rearrangements, deduced from the studies presented by the sequence of journal articles[16-29] and summarized in the book[13].

Rearrangements in a solid state are a crystal growth by nucleation and propagation of interfaces. Neither ferromagnetic and ferroelectric phase transitions, nor phase transitions involving the orientation-disorder crystal (ODC) phase are excluded from this rule. Not a single sufficiently documented example exists of a transition being homogeneous (cooperative).

The nuclei are located in specific crystal defects - microcavities of a certain optimum size. These defects contain information on the condition (*e.g.*, temperature) of their activation and orientation of the resultant crystal lattice. The nucleation can be epitaxial, in which case a certain orientation relationship between the initial and resultant structures is observed.

The interface is a rational crystallographic plane of the resultant crystal lattice. It is named "contact interface" owing to a direct molecular contact between the two lattices without any intermediate layer. The molecular rearrangement proceeds according to *edgewise* (or *stepwise*) mechanism (Fig.1) involving formation of "kinks" (steps) at the flat interface and filling them, molecule-by-molecule, until the layer is complete, and building successive layers in this manner.

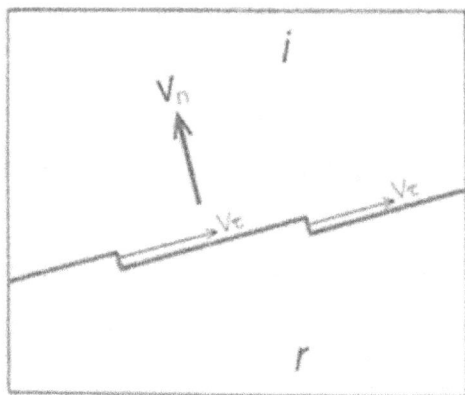

**Figure 1.** The *edgewise* mechanism of phase transitions and any other rearrangements in solid state, such as at domain boundaries. The sketch illustrates the mode of advancement of interface in the **n** direction by shuttle-like strokes of small steps (kinks), filled by molecule-by-molecule, in the **t** direction; *i* and *r* – are initial and resultant crystals, respectively. (A crystal growth from liquids is realized by the same manner). The kinks may consist of a single molecular layer or be a ladder-like conglomeration of smaller steps. Refer to [24,13] for more detailed description

# 8. Accounting for Ferromagnetism and its Manifestations (Including Problems Cited by Feynman)

This will be done below within reasonable limits of a single article - mostly in a synopsis form.

### 8.1. Some Problems Are Automatically Eliminated

There are two types of ferromagnetic phase transitions - second order and first order. (Only one exists).

Application of the statistical mechanics to first-order ferromagnetic phase transitions. (Not applicable).

The Curie point is blurred and subjected to hysteresis. (Phase transition temperature is not a Curie point).

Magnetocrystalline (anisotropy) energy. (The notion is eliminated, considering that spin orientation is *fully determined* by the crystal structure).

### 8.2. Stability of a Ferromagnetic State. (Feynman: "Why Ferromagnetism Exists at All?")

Ferromagnetic state is a "slave" of crystal structure. A particular spin alignment ("magnetic structure") is determined by the requirements of crystal packing. The magnetic structure is an element of that 3D packing, contributing a small positive or negative addition to the total crystal free energy. Ferromagnetism materializes in those cases when minimum free energy of the *crystal packing* requires placing spin carriers in the positions with their spins not mutually compensated. Despite of the possible destabilizing effect of the magnetic interaction, it is too weak to make any alternative crystal structure preferable. In brief: contribution of the magnetic interaction to the total crystal free energy is small as compared to that of crystal bonding; a ferromagnetic crystal is stable due to its low *total* free energy *in spite* of the possible destabilizing effect of the magnetic interaction.

### 8.3. Feynman: "Why Is a Piece of Lodestone in the Ground Magnetized?"

By razing this question, Feynman meant that, besides the stability problem, there must be an original cause turning non-ferromagnetic lodestone to ferromagnetic. Answer: it became ferromagnetic in the prehistoric times during its crystallization from liquid phase. The ferromagnetic state of lodestone is an inherent element of its crystal structure.

### 8.4. Existence of a Great Variety of Non-Collinear Magnetic Structures

These are some types of magnetic structures in crystals: "simple ferromagnetic", "simple antiferromagnetic", "ferrimagnetic", "weakly ferromagnetic", "weakly non-collinear antiferromagnetic", "triangle", "simple helical", "ferromagnetic helical", and more. Only in the heavy metallic rare earths the following magnetic structures were listed[9]: "ferromagnet", "helix", "cone", "antiphase cone", "sinusoidally modulated", "squarewave modulated". The diversity in the mutual positions and orientations of spins can only be matched by the diversity in the world of crystal structures. This is not accidental: a magnetic structure is *imposed* by the crystal, being secondary to the requirements of the crystal geometry.

### 8.5. Paramagnetic Phase

It is usually assumed, as Weiss did, that the magnetic dipoles of the high-temperature phase of a ferromagnet lost their ferromagnetic alignment due to thermal rotation. The Weiss' view is understandable, for in his times the orientation-disordered crystals (ODC) were not yet discovered. The atoms and molecules, and not their spins alone, in the ODC state are engaged in a hindered thermal rotation. Besides, a zero magnetic moment of the high-temperature phase in question can also be not owing to the ODC state, but due to mutual compensation of its spins in the centrocymmetrical structure.

## 8.6. Ferromagnetic Phase Transitions

Reorientation of spins involved in these phase transitions requires changing the orientation of spin carriers. The only way to achieve that is replacing the crystal structure. This occurs by nucleation and interface propagation. It follows that *all* ferromagnetic phase transitions without exceptions are "magnetostructural". The term, however, is defective in the sense that it suggests existence of ferromagnetic phase transitions without structural change.

## 8.7. Magnetization by Interface Propagation

The conventional theory does not explain why magnetization occurs in this manner rather than cooperatively in the bulk. Once again: magnetization is not a spin reorientation in the same crystal structure, but requires turning the spin carriers. The only way to turn the carriers is by crystal rearrangement. The mechanism of crystal rearrangements is nucleation and interface propagation. The possibility of a cooperative magnetization "by rotation" is thus ruled out[13,31].

## 8.8. Magnetization "Switching" and "Reversal"

Their experimentally estimated ultimate speed in single-domain particles turned out three orders of magnitude lower than theoretically predicted[30]. The cause: whether they are activated by temperature, pressure, or external magnetic field, they always materialize by a relatively slow process of nucleation and propagation of interfaces[31,32].

## 8.9. The Origin of Magnetic Hysteresis

The current theory was powerless to deal with magnetic hysteresis other than in a phenomenological manner, while its physical cause remained a question mark. *Solution*: Magnetic hysteresis is a reflection of the *structural* hysteresis both in ferromagnetic phase transitions and in magnetization of domain systems. They require 3-D nucleation to begin and 2-D nucleation to proceed. The nucleation is heterogeneous, localized in specific defects – microcavities – where nucleation lags are encoded. These nucleation lags are the cause of magnetic hysteresis[13,32].

## 8.10. Formation of Magnetic Hysteresis Loops

The "sigmoid" shape of the hysteresis loops is due to the balance between the increase in nucleation sites per volume unit and the decrease in the amount of the original phase

## 8.11. Specific Heat near the Curie Transition

(Feynman: "One of the challenges of theoretical physics today is to find an exact theoretical description of the character of the specific heat near the Curie transition - an intriguing problem which has not yet been solved. Naturally, this problem is very closely related to the shape of the magnetization curve in the same region").

The cooperative "Curie transition" does not exist. Solid-state phase transitions occur by nucleation and growth

(Section 6). What believed to be a specific heat anomaly (called $\lambda$ -anomaly) is not anomaly at all. It is the *latent heat* of a first-order phase transition (Fig. 2). Refer to[33] and Chapter 3 in[13].

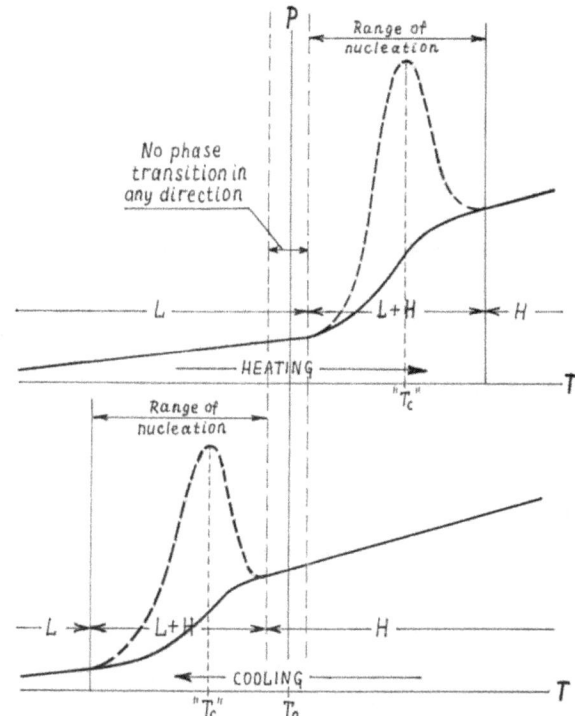

**Figure 2.** The "anomalous" peaks of a physical property P, believe to be a heat capacity or magnetization, reside in the ranges of transition (actually, ranges of nucleation). The "critical (Curie) point $T_c$" at the l-peak top (the common choice) is a subject of hysteresis, for there are two non-overlapping transition ranges, one above $T_o$ - for heating, and the other below $T_o$ - for cooling. In the adiabatic calorimetry these peaks are not a specific heat, but the *latent heat* of first-order (nucleation and growth) phase transitions. A differential scanning calorimetry would reveal the peak in a cooling run actually looking downward, being exothermic

## 8.12. Ferromagnetic Domain Structure

An essential fact regarding ferromagnetic domain structure is that it is not specifically rooted in a ferromagnetic state, as Landau and Lifshitz[34] assumed. Domain structures are found also in antiferromagnetics, ferroelectrics, superconductors, organic crystals, etc. Their origin is *structural*. A ferromagnetic domain structure originates by multiple nucleation of the ferromagnetic phase in several equivalent structural orientations within the paramagnetic matrix. Growth of these nuclei and subsequent "magnetic aging" proceed toward minimizing the magnetic energy. Refer to[13], Sec. 2.8.6, 4.5 and 4.9.

## 8.13. Barkhausen Effect

The effect-short advances and stops during magnetization process - is foreign to the traditional theory. The exchange field theory did not assume it. The domain theory may account only for the largest magnetization jumps, but they always consist of much smaller steps. The recent scientific work was devoted only to the phenomenological description of the effect, shedding no light on its nature[35]. But the effect is a direct manifestation of the crystal growth. In order to lower the crystal free energy in the applied magnetic field $H$, the spins of the ferromagnetic crystal have to turn toward the $H$ direction, causing the structural rearrangement at the interfaces as shown in Fig. 1. Quick recrystallization of a whole layer at the domain boundary produces a magnetic "jump". The rearrangement of every successive layer is delayed by availability of next nucleus. The layers can be as thin as one lattice space, or they can be conglomerations of numerous elementary layers. In the latter case larger steps ("avalanches") appear on the magnetization curve. A quick restructuring of a whole domain would produce the largest step, but it will inevitably consist of many smaller ones. Refer to[13], Sec. 4.10 and Addendum H.

### 8.14. Magnetostriction of Fe

The phenomenon is not a kind of deformation, as usually believed. The α-Fe has a tetragonal rather than a cubic crystal structure. The magnetostriction results from the structural rearrangement, induced by application of magnetic field, that makes the direction of the longer crystallographic axis of the participated domains coincide with, or become closer to, the direction of the applied magnetic field[13,36].

### 8.15. Magnetocaloric Effect

It was acknowledged[37] that the "underlying physics behind the magnetocaloric effect is not yet completely understood". Now the physical nature of a "giant" magnetocaloric effect is explained in terms of the new fundamentals of phase transitions, ferromagnetism and ferroelectricity[13]. It is the *latent heat* of structural (nucleation-and- growth) phase transitions from a normal crystal state to the orientation-disordered crystal (ODC) state where the constituent particles are engaged in thermal rotation. The ferromagnetism of the material provides the capability to *trigger* the structural phase transition by application of magnetic field[38].

### 8.16. Disparity with Ferroelectricity

Ferromagnetism and ferroelectricity are very similar phenomena with analogous set of manifestations. The standard theory was unable to find a unified approach to them since the Weiss/Heisenberg molecular field was applied only to ferromagnetism. No analog to it was found (or even needed) for ferroelectricity. Solution: This profound inconsistency disappears after the Weiss/Heisenberg molecular field is eliminated from consideration. Now the two phenomena have quite parallel explanations[13].

---

# REFERENCES

[1]    F. Seitz, The Modern Theory of Solids, Mc Grow-Hill (1940), or any of the numerous subsequent editions.

[2]    I. E. Tamm, Fundamentals of the Theory of Electricity, Mir Publications, Moscow (1979).

[3]    R. M. Bozorth, Ferromagnetism, D. Van Nostrand, New York (1951).

[4]    S. V. Vonsovskii, Magnetism, vol. 2, Wiley (1974).

[5]    W. Heisenberg, 1928, Zur theorie des ferromagnetismus, Z. Physik, 49, 619-636.

[6]    C. Kittel, Introduction to Solid State Physics, 4th Ed., Wiley, (1971).

[7]    B. I. Bleaney, B. Bleaney, Electricity and Magnetism, Oxford, Clarendon Press (1963).

[8]    K. P. Belov, Magnetic Transitions, Boston Tech. Publ. (1965).

[9]    J. Crangle, The Magnetic Properties of Solids, Edward Arnold, London (1977).

[10]   R. P. Feynman, R. B. Leighton, M. Sands, The Feynman Lectures on Physics, v.2, Addison-Wesley (1964).

[11]   J. H. Samson, 1995, Classical effective Hamiltonians, Wigner functions, and the sign problem, Phys. Rev., B 51, 223-233.

[12]   R. S. Preston, 1968, Temperature dependence of isomer shift and hyperfine field near the Curie point in iron, J. Appl. Phys., 39, 1231.

[13]   Y. Mnyukh, Fundamentals of Solid-State Phase Transitions, Ferromagnetism and Ferroelectricity, Authorhouse, 2001[or 2nd (2010) Edition].

[14]   Sen Yang , Xiaobing Ren, Xiaoping Song, 2008, Evidence for first-order nature of the ferromagnetic transition in Ni, Fe, Co, and CoFe2O4, Phys. Rev., B 78,174427.

[15]   Y. Mnyukh, Second-order phase transitions, L. Landau and his successors, http://arxiv.org/abs/1102.1085.

[16]   Y. Mnyukh, 1963, Laws of phase transformations in a series of normal paraffins, J. Phys. Chem. Solids, 24, 631-640.

[17]   A. I. Kitaigorodskii, Y. Mnyukh, Y. G. Asadov, 1963, The polymorphic single crystal to single crystal transformation in p-dichlorobenzene, Soviet Physics - Doclady, 8, 127- 130.

[18]   A. I. Kitaigorodskii, Y. Mnyukh, Y. Asadov, 1965, Relationships for single crystal growth during polymorphic transformation, J. Phys. Chem. Solids, 26, 463-472.

[19]   Y. Mnyukh, N. N. Petropavlov, A. I. Kitaigorodskii, 1966, Laminar growth of crystals during polymorphic transformation, Soviet Physics - Doclady, 11, 4-7.

[20]   Y. Mnyukh, N. I. Muse, A. I. Kitaigorodskii, 1967, Crystal growth in polymorphic transitions of glutaric acid hexachloroethane, Soviet Physics - Doclady, 12, 409-412.

[21]   Y. Mnyukh, N. I. Musaev, 1969, Mechanism of polymorphic transition from the crystalline to the rotational state, Soviet Physics - Doclady, 13, 630-633.

[22]   Y. Mnyukh, 1972, Molecular mechanism of polymorphic transitions, Soviet Physics - Doclady, 16, 977-980.

[23]   Y. Mnyukh, N. N. Petropavlov, 1972, Polymorphic transitions in molecular crystals-1. Orientations of lattices and interfaces, J. Phys. Chem. Solids, 33, 2079-2087.

[24] Y. Mnyukh, N. A. Panfilova, 1973, Polymorphic transitions in molecular crystals-2. Mechanism of molecular rearrangement at 'contact' interface, J. Phys. Chem. Solids, 34, 159-170.

[25] Y. Mnyukh, N. A. Panfilova, 1975, Nucleation in a single crystal, Soviet Physics - Doclady, 20, 344-347.

[26] Y. Mnyukh, N. A. Panfilova, N. N. Petropavlov, N. S. Uchvatova, 1975, Polymorphic transitions in molecular crystals - 3. Transitions exhibiting unusual behavior, J. Phys. Chem. Solids, 36, 127-144..

[27] Y. Mnyukh, 1976, Polymorphic transitions in crystals: nucleation, J. Crystal Growth, 32, 371-377.

[28] Y. Mnyukh, 1979, Molecular mechanism of polymorphic transitions, Mol. Cryst. Liq. Cryst., 52, 163-200.

[29] Y. Mnyukh, 1979, Polymorphic transitions in crystals: Mol. Cryst. Liq. Cryst., 52, 201-218.

[30] I. Tudosa, C. Stamm, A. B. Kashuba, F. King, H. C. Siegmann, J. Stöhr, G. Ju, B. Lu, D. Weller., 2004, The ultimate speed of magnetic switching in granular recording media, Nature, 428, 831-833.

[31] Y. Mnyukh, The physics of magnetization, http://arxiv.org/abs/1101.1249.

[32] Y. Mnyukh, Hysteresis and nucleation in condensed matter, http://arxiv.org/abs/1103.2194.

[33] Y. Mnyukh, Lambda- and Schottky-anomalies in solid-state phase transitions, http://arxiv.org/abs/1104.4637.

[34] Collected Papers of L.D. Landau, Gordon & Breach (1967).

[35] G. Durin, S. Zapperi, The Barkhausen effect, cond-mat/0404512.

[36] Y. Mnyukh, The true cause of magnetostriction, http://arxiv.org/abs/1103.4527.

[37] N. A. de Oliveira, P. J. von Ranke, 2010, Theoretical aspects of the magnetocaloric effect, Phys. Reports, 489, 89-159.

[38] V. J. Vodyanoy, Y. Mnyukh, The physical nature of "giant" magnetocaloric and electrocaloric effects, http://arxiv.org/abs/1012.0967.

# Magnetization of Ferromagnets

A detailed new explanation of spontaneous magnetization, domain structure and magnetization process is presented. A very realistic assumption that *spin direction is fixed in the structure of its atomic carrier* easily solves all problems of the current theory of ferromagnetism, eliminating such its attributes as Heisenberg theory of ferromagnetism, anisotropy energy and Bloch wall. The source of spontaneous magnetization is *crystal field* that sets up the orientations of crystal particles. Thus, magnetization (spin reorientation) can materialize only by structural rearrangements, but not by spin rotation in the same structure. It proceeds by propagation of *contact* interfaces where magnetic particles are relocated one particle at a time in accordance with the *stepwise* mechanism of solid-state reactions.

American Journal of Condensed Matter Physics 2014, 4(4): 78-85
DOI: 10.5923/j.ajcmp.20140404.03

# Magnetization of Ferromagnets

## Yuri Mnyukh

76 Peggy Lane, Farmington, USA

**Abstract**  A detailed new explanation of spontaneous magnetization, domain structure and magnetization process is presented. A very realistic assumption that *spin direction is fixed in the structure of its atomic carrier* easily solves all problems of the current theory of ferromagnetism, eliminating such its attributes as Heisenberg theory of ferromagnetism, anisotropy energy and Bloch wall. The source of spontaneous magnetization is *crystal field* that sets up the orientations of crystal particles. Thus, magnetization (spin reorientation) can materialize only by structural rearrangements, but not by spin rotation in the same structure. It proceeds by propagation of *contact* interfaces where magnetic particles are relocated one particle at a time in accordance with the *stepwise* mechanism of solid-state reactions.

**Keywords**  Magnetization, Ferromagnetism, Domain structure, Domain boundary, Bloch wall, Spin rotation, Heisenberg theory, Anisotropy energy, Phase transitions

## 1. Introduction

The *cause of spontaneous magnetization* and the *process of magnetization* are two major problems of the theory of ferromagnetism. They are not solved correctly by the conventional theory. Their new solution has been put forward in relation to other issues [1-6], but did not attract due attention. The purpose of this article is to present the topic on its own in a consistent and detailed form in contraposition with the conventional theory. The latter is usually described as complete and impeccable. The book by Kittel [7] is a good example. His account will be used now to identify what is erroneous in it and to spot blanks. Other source for that matter is the book by Bozorth [8]. Some figures and arguments from our previous articles will be repeated for the sake of entirety.

The revised molecular mechanism of magnetization has come about from our experimental study and analysis of solid-state phase transitions as a function of temperature. An essential kind of them is *ferromagnetic* phase transitions, frequently referred to as "ferromagnetic ordering" in literature. Indeed, a ferromagnetic phase transition is a *magnetization process* with changing of the crystal structure. As far as its molecular mechanism is concern, the type of driving force (temperature) is not essential, considering that the same phase transitions could be driven by magnetic field as well. It turns out, a magnetization process with or without crystallographic change is based on the same principle, which is *rearrangement of the crystal*.

* Corresponding author:
yuri@mnyukh.com (Yuri Mnyukh)
Published online at http://journal.sapub.org/ajcmp

## 2. Solid-State Structural Rearrangements

The following is the general mechanism of structural rearrangements, deduced from the studies of phase transitions, presented initially by a sequence of journal articles [9-20] and then summarized in the book [1] and later by a set of articles [4-6, 21, 22].

- Rearrangements in a solid state materialize by a crystal growth involving nucleation and propagation of *contact* interfaces. Neither ferromagnetic phase transitions, nor ferroelectric phase transitions are excluded from this rule. Not a single sufficiently documented example exists of this process being "second order" (homogeneous, cooperative) suggested by the Landau theory.

- The nucleation differs greatly from its classical theory. It makes hysteresis inevitable. The nuclei are located in specific crystal defects − microcavities of a certain optimum size. These defects contain individual information on the condition needed to activate them (*e.g.*, temperature, applied magnetic field) and the orientation of the resultant crystal.

- The nucleation can be *epitaxial*, in which case a certain orientation relationship between the initial and resultant structures is observed. It occurs in prominently layered crystals or when lattice parameters of the phases are very close. In strong epitaxial cases the *contact* interface becomes a *twin* interface.

- The interface is a rational crystallographic plane of the resultant crystal lattice. It is named *contact interface* owing to direct molecular contact between the two lattices without any intermediate layer. The molecular rearrangement proceeds according to *stepwise* (or *edgewise*) mechanism (Fig.1) consisting of formation of "kinks" (steps) at the flat interface and filling them, molecule-by-molecule, until the layer is complete, and building successive layers in this manner.

# 3. Conventional Standing on Magnetization

The following brief description of the conventional views on magnetization involves some representative excerptions from the Kittel book [7], clarifications in square brackets and comments.

"Given an internal [spin] interaction tending to line up the magnetic moments parallel to each other, we shall have a ferromagnet. Let us postulate such an interaction and call it the **exchange field.** The orienting effect of the exchange field is opposed by thermal agitation, and at elevated temperatures the spin order is destroyed."

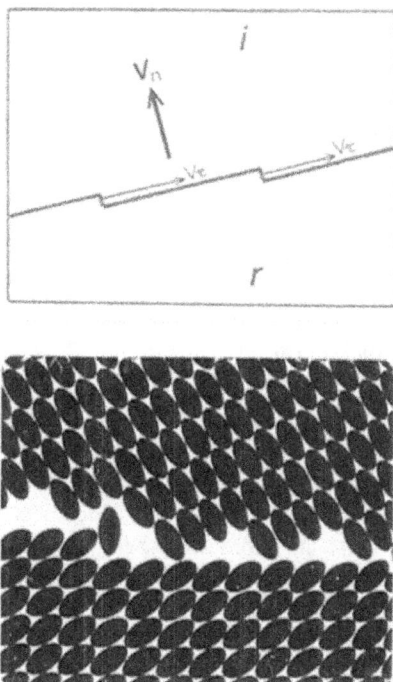

(Above): A sketch illustrating the mode of advancement of interface in the **n** direction by shuttle-like strokes of small steps (kinks), filled by molecule-by-molecule, in the **τ** direction; *i* and *r* – are initial and resultant crystals, respectively. (Crystal growth from liquids is realized by the same manner).

(Below): Molecular model of phase transition in a crystal. The *contact* interface is a rational crystal plane in the resultant phase, but not necessarily in the initial (upper) phase. The interface advancement proceeds by molecule-by-molecule filling kinks and then layer-by-layer in this manner. The gap of 0.5 molecular layer (on average) is wide enough to provide steric freedom for the molecular relocation at the kink, but is sufficiently narrow for the relocation to occur under attraction from the side of resultant crystal.

**Figure 1.** The *stepwise* (or *edgewise*) mechanism of phase transitions and any other rearrangements in solid state

(1) The quantum-mechanical theory of spontaneous magnetization by Heisenberg was accepted (without attribution to its author) as the overall approach. (2) If the

theory was valid, why was the *exchange field postulated*? (3) It was also postulated, implicitly, that the system of magnetic moments was an independent entity rather than a property of the crystal. In such a case, *spin interaction* could be the only conceivable cause of the spontaneous magnetization.

"Electronic magnetic moments of a ferromagnet are essentially all ["essentially all" or all?] lined up when regarded on a microscopic scale" [of crystal unit cells, Fig.2].

"Actual specimens [*polydomain* crystals that implied to be structural *single crystals*] are composed of a number of small regions called domains, within each of which the local magnetization is saturated. The directions of magnetization of different domains, however, need not necessarily be parallel" [Fig. 3]. The resultant magnetic moment of a polydomain crystal can be far from saturation − down to zero.

"The increase in the magnetic moment of the specimen under the action of an applied magnetic field takes place by two independent processes: (1) in weak applied fields the volume of domains which are favorably oriented with respect to the field increases at the expense of unfavorably oriented domains; (2) in strong applied fields the magnetization rotates toward the direction of the field".

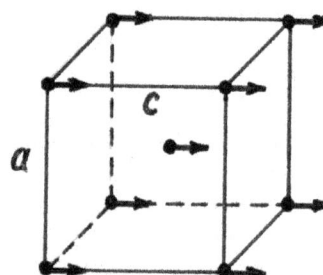

**Figure 2.** Unit cell of ferromagnetic α-Fe crystal and its spontaneous magnetization.. The difference *a* (= *b*) - *c* is very small. Only recently it was shown that it is tetragonal rather than truly cubic

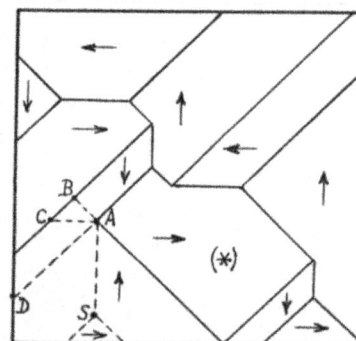

**Figure 3.** A schematic illustrating the features of domain patterns in the Fe polydomain crystals (shown is (*100*) plane). The broken lines help to comprehend their "rules": only interface AS is correct, not AB, AC or AD

The direction of spontaneous magnetization is called "direction of easy magnetization". The term is ambiguous in respect to an isolated domain: it is already spontaneously magnetized to saturation $M_S$ in that direction. Evidently, it allows a possibility for the directions of spontaneous magnetization to be deviated from their "normal" position. The term acquires clear meaning when dealing with a polydomain crystal.  Achieving $M_S$ (by propagation of the domain boundaries until their disappearance), would require the lowest strength of the applied field $H$ when it is parallel to the spontaneous magnetization of the arising crystal.

Other crystallographic directions are "hard directions" of various degrees of "hardness". In Fe, in order to magnetize a polydomain crystal in the $a$ direction when $H$ is not along $a$, the $H$ should be stronger, since its "driving" strength is reduced to $H = |H \cdot cos (H \char"5E a)|$. Then, to achieve $M_S$ to be along $H$, the $H$ strength should be even higher. In that case magnetization in the $a$ direction has been completed; there are no domain boundaries any more in the polydomain crystal. It must be remagnetized from the "easy" to a "not easy" direction − and the conventional theory does not see any alternative to a "magnetization rotation" in the crystal. The direction of spontaneous magnetization is deemed to be only "preferable", subject of changing by applied magnetic field against resistance of the crystal. A special phenomenon was introduced:

"There is an energy in a ferromagnetic crystal which directs the magnetization along certain definite crystallographic axes called directions of easy magnetization. This energy is called the **magnetocrystalline** or **anisotropy energy**."

Like in the case of the "exchange field", existence of that energy was also *postulated*, for its physical nature was not specified. Whether it is considered independent of the "exchange field" remains murky, but possibly not, as the two are cited to be "various contributions to the energy". Thus, the standard theory of magnetization rests on two *postulated* energies: of the "exchange field", arranging spins in parallel, and the "anisotropy", directing them in the crystal lattice.

The microscopic picture of domain boundaries is currently represented by a "Bloch wall" (Fig. 4), again involving the idea of spin rotation in the crystal.

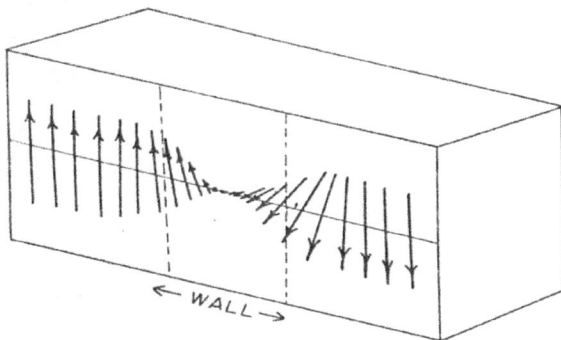

**Figure 4.**   Theoretical 180° - "Bloch wall" in a single-crystal continuum. It takes   many lattice periods for the spins to gradually turn to the opposite direction. There are also "walls" of different angles of dipole rotation, e.g. "90°-walls"

"A **Bloch wall** in a [single] crystal is the transition layer which separates adjacent regions (domains) magnetized in different directions.  The entire change in spin direction between domains does not occur in one discontinuous jump across a single atomic plain, but takes place in a gradual way over many atomic planes … In iron the thickness of the transition region is about 300 lattice constants".

Last major point in this brief description of the standard views on magnetization is *origin* of the domains. The phenomenon is claimed to be "a natural consequence of the various contributions to the energy − exchange, anisotropy, and magnetic − of a ferromagnetic body", and accompanied by a sketch [not shown] to illustrate that magnetic energy of a ferromagnetic single crystal is lowered by its separation into domains. There were no attempts to explain formation of the intricate domain structure like pictured in Fig. 3.

## 4. Questions Waiting for Answers

The Heisenberg's theory has been accepted owing to high prestige of its author and apparent lack of an alternative. There was, however, an authoritative skeptical voice. In contrast to Kittel and almost all other authors, presentation of ferromagnetism by R. P. Feynman [23] contained discussions of its serious weaknesses and inconsistencies. It was emphasized that while the quantum-mechanical theory by Heisenberg intended to explain the phenomenon of spontaneous magnetization by extremely strong spin interaction due to overlapping electron shells, it consistently produced the wrong sign of that interaction. Feynman concluded: "To the theoretical physicists, ferromagnetism presents a number of very interesting, unsolved, and beautiful challenges".

In the framework of the magnetization problem only, here are the major questions waiting for answers, the first one belonging to Feynman:

- "We still have the question: why is a piece of lodestone in the ground magnetized?".
- Is the cause of spontaneous magnetization the Heisenberg's electron exchange field?
- What makes spontaneously magnetized matter thermodynamically stable?
- What is the physical nature of the anisotropy (or magnetocrystalline) energy?
- How the anisotropy energy chooses and aims the spontaneous magnetization in one or another direction?
- Why does magnetization by magnetic field proceed by formation and advancement of domain boundaries, rather than by (cooperative) turning all spins simultaneously?
- Why do domain boundaries move slowly rather than by speed of magnetic waive?
- When magnetization is claimed to be "by rotation" of spins, is the rotation a cooperative (all spins together)?
- How is the peculiar domain patterns like shown in Fig. 3 formed?
- Why are the domain boundaries straight?

- How can the "Bloch wall" explain the "head to tail" attraction of elementary magnets at the domain boundaries if it is so wide as 300 lattice periods?
- How does the "Bloch wall" cover the multiple cases when the system to be magnetized is antiferromagnetic, helical or with spins being in a state of thermal agitation?

Before entering into dismantling of the standard theory, its correct part is to be recognized. That part is a description of the first magnetization stage, which is a movement of the domain boundaries until the whole polydomain crystal acquires the spin orientation of the domain having the orientation closest to (not necessarily coincident with) the *H* direction. All the rest − structure of polydomain crystals, origin of spontaneous magnetization, spin rotation, Bloch wall, anisotropy energy − will be challenged and replaced in the sections that follow. The first step in that direction is to reveal that the Heisenberg's theory of electron-exchange field is not valid.

## 5. Theoretical Predicament Ferromagnetism Cannot Exist

The Heisenberg's theory was an attempt to find the cause of the phenomenon of ferromagnetism narrowly interpreted as a thermodynamically stable existence of parallel spins in crystals − like that in Fig. 2. It had to be explained how that state can be stable up to the temperatures so high as 1042 K in Fe. It seemed unavoidable to conclude (incorrectly) that the force holding the elementary dipoles in parallel is their mutual interaction. In such a case, this force had to be about $10^4$ times stronger than their magnetic interaction alone. Heisenberg accepted that as a fact and developed its quantum-mechanical interpretation: it is overlapping of the electron shells that results in extremely strong *electron exchange interaction*, sufficient for the stable collinear spin alignment up to the observed high temperatures.

General acceptance of the Heisenberg's theory remains unshakable to the present days. The contradictions, blank areas, and vast disagreements with experiment are being ignored. Yet, there are exceptions, listed in [24], where those inconsistencies were noted. They led Feynman to state "One challenge is to understand why [ferromagnetism] exists at all". In other words, why a spontaneously magnetized state is stable?

In fact, the theory had to be rejected from the very beginning: it produced a *wrong sign* of the exchange forces. This means that the apparent electron-exchange field, if existed, would produce a spin *repulsion* of the same great strength as the attraction it had to prove. The repeated calculations consistently produced the same result, actually meant that ferromagnetism must not exist. The sign problem was carefully examined [25] and found fundamentally unavoidable in the Heisenberg model. It was suggested that the "neglect of the sign may hide important physics." (This subject is described in [4] in more detail).

The theory of electron-exchange field should be nullified not only due to its failures; it was not needed at all for explanation of ferromagnetism / spontaneous magnetization. It will be shown below that the parallel alignment of spins is set up by *crystal forces*, and not by spin interaction.

## 6. Tearing Down the Bloch Wall

The "Bloch wall" (Fig. 4) was a derivative of two erroneous theories, namely, the molecular / exchange field and anisotropy energy. This type of a boundary between ferromagnetic domains was not discovered experimentally, neither any credible circumstantial evidence existed. It was invented to help the Heisenberg's theory overcome a predicament: prohibitively high energy of the domain boundaries. In terms of the theory, the exchange energy between neighboring magnetic dipoles on the opposite sides of a 180° domain boundary (i.e., when the dipoles are antiparallel) is at its maximum. A gradual change of the dipole directions over a number of lattice periods would lower that energy, but their deviation from the crystallographic "direction of easy magnetization" would have the opposite effect. The "Bloch wall" was the result of a theoretical optimization of the "exchange field" and "anisotropy" energies effects. This did not mean that the "Bloch wall" was a low energy boundary: it was only minimal under the above theoretical assumptions.

Some calculations have set the "wall" thickness from 50 to 200 nm in Fe. Kittel came up with 300 lattice spacings (~86 nm). According to Hubert [26], it is "several thousands" crystal spacings. Understanding "several" as minimum of three, this gives rise to the walls in Fe at least ten times wider than stated by Kittel. There have been no experimental data validating these numbers.

It will be revealing to compare the ~86 nm width of the "Bloch wall" with the size of domains in Fe inferred also from the standard theory. Bozorth's estimation [8], in agreement with other authors, was ~10 nm for the maximum size of a Fe crystal remaining as a single domain. This means that a larger crystal, for example, 12 nm will, or at least can, consist of two domains. But Bozorth warned us not to take the estimate "too literally". Therefore, we assume that not 12, but a 120 nm particle can consist of two, and for better illustration, a 180 nm particle of three domains. Comparison of the result with the 86 nm "Bloch walls" between them is shown in Fig. 5.

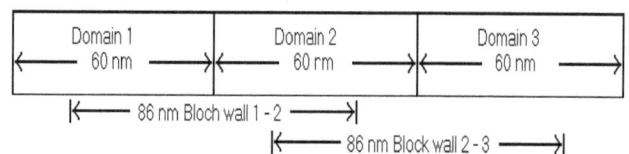

**Figure 5.** Inconsistence between estimates of the "Bloch wall" thickness and the dimensions of ferromagnetic domains by the theory

The result is nonsensical. Even changing one of the values under comparison by another order of magnitude (e.g., increasing the three-domain particle to 1800 nm) would not produce a satisfactory result, for even then as much as 14% of every domain will be occupied by the "walls".

As usual with artificial constructions, every particular case required a "wall" modification. For example, for the angles 180°, 109.47° and 70.53° at the domain boundaries in Ni, one has to assume three different types of the "walls". What's more, the "Bloch wall" was initially designed specifically for ferromagnetics, and not antiferromagnetics. The theory of "domain walls" became so cumbersome that an entire book [26] was needed to overview it. Unfortunately, wrong premises made that work useless.

# 7. Polydomain Crystal

## 7.1. Not a Single Crystal

The "Bloch wall" was implied to be a purely magnetic phenomenon residing in the *single crystal medium*, and the same idea was hidden in all relevant theories. It could even be supported by X-ray single-crystal patterns of polydomain Fe crystals. Yet, the Fe polydomain crystals are only *pseudo*-single crystals. They are conglomerations of single crystals (domains) of identical crystal structure differently oriented relative to one another.

During all 20th century the structure of iron at room temperature believed to be a truly cubic. But it is shown now [1, 27] that Fe unit cell is a tetragonal pseudo-cubic, with a minute difference $(a = b) - c$. The arrows in Figs. 2 and 3, indicating the directions of spins, represent the directions of the lattice parameter $c$ as well. This brings about an important revelation about a domain boundary. It is not simply a place where spins change their direction; it is the contacting place of two individual crystal lattices. While the structures of the contacting domains are identical, their different spatial orientation makes their boundary no physically different from those in phase transitions. The domain boundaries are *contact interfaces*, not "Bloch walls".

## 7.2. Formation from Paramagnetic Phase

In order to describe formation of polydomain structures of the patterns as elaborate as in Fig. 3, the general mechanism of solid-state reactions outlined in Section 2 had to be invoked. That is why their formation has not been explained up to now. They are formed by epitaxial nucleation and growth.

Common solid iron is polycrystalline, consisting of randomly oriented grains separated by irregular boundaries. Every grain has a polydomain structure. Each domain is a single crystal with its own direction of the elementary magnetic dipoles (spins) along the crystal parameter $c$.

Let us trace the metamorphoses every grain undergoes during its *paramagnetic → ferromagnetic* phase transition at 770℃. Above that temperature it is a single crystal of cubic symmetry where atoms and their spins are disoriented due to thermal agitation. It is an "orientation-disordered crystal"

(ODC), known in the non-magnetic world as well. Below that temperature the phase transition proceeds by multiple nucleation and growth of the normal solid phase. Because the nucleation is *epitaxial*, and the matrix is cubic, the nuclei appear arbitrarily in any one of six orientations (Fig. 6). When the crystals growing from those nuclei meet, there is one chance that their spins will be in parallel ("0°-boundary"), one chance - antiparallel ("180°-boundary"), and four chances - "90°-boundary". It is probable that magnetism in that phase transition comes to play only at the stage when the growing single crystals come very close to each other, so that their spins can start interacting; until then the ODC-phase between them will have a screening effect. When they ultimately meet, *i.e.*, when the phase transition is completed, the initial domain interfaces, if not quite well organized, will be finalized already in solid state by the *magnetic recrystallization* [1] according to the general mechanism of all solid-state reactions (Fig. 1). It will take place to minimize free energy of the interfaces. This involves two requirements: best possible crystal packing and mutual attraction of the magnetic dipoles. The case under consideration is such that both requirements can be well satisfied: the crystal lattices on both sides will match almost perfectly to become a *twin interface of zero thickness* and, simultaneously, the dipoles will be arranged in a "head-to-tail" order. That is how the domain structures like one in Fig. 3 are formed.

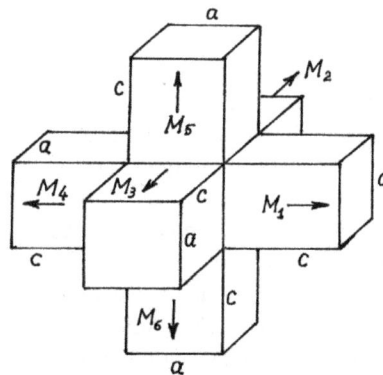

**Figure 6.** Six mutual directions of spontaneous magnetization $M_s$ that make 90° and 180° to one another in a polydomain iron crystal. While the unit cell dimensions $c$ and $a = b$ are very close, it is axis $c$ which is the $M_s$ direction. (The figure is not intended to illustrate domain boundaries)

# 8. Domain Interface at Molecular Level

The first observation of the stepwise (or edgewise) mode of interface propagation in a solid-state phase transition, illustrated by sketch in Fig. 1, was reported in 1966 [10]. A consonance with a crystal growth from liquid phase was specified. It was concluded that the phenomenon is a *crystal growth* in solid phase, in which successive crystal layers are built up by filling steps ["kinks"] one molecule at a time. The observations, however, were made only with the optical resolution not better than ~100 nm achievable at that time. It was stated that "while our technique did not allow us to observe elementary steps of growth, their existence is undoubted".

Experimental proof of that prediction had to wait for 37 years until Novoselov *et al.* (NGDHG) [28] recorded a domain "wall" with a high resolution attained never before. They did not see the "Bloch wall of 300 lattice periods wide", but found that the ferromagnetic domain interface propagated "via clear quantized jumps matching the lattice periodicity. The distance between the equivalent sites was measured to be 1.6 $^+$ 0.2 nm". The authors interpreted the interface movements as following the Peierls potential of the crystal lattice and believed that "the presence of a periodic atomic landscape [is] impeding movements of domain wall".

Also observed were "kinks" running along the interface. That led to speculations why the "wall" does not jump as a whole between the neighboring minimums of the Peierls potential. As a result, NGDHG stated that "further theoretical and experimental work is required to understand atomic scale [domain walls] dynamics. They did not recognize that *dealt with magnetic particles* rather than spins alone. The *stepwise* nucleation-and-growth mechanism of solid-state reactions explains all their observations.

## 9. Assumption: Spins Uniquely Oriented in the Structure of Their Atoms

Possibility of spin rotation within magnetic atoms is invisibly (and subconsciously) present in the current theory. It is hidden in the claims (*e.g.*, by Bozorth [8]) that spontaneous magnetization $M_s$ in the same crystal can be rotated by application of magnetic field $H$.

Let us turn to an atom of Fe. In the crystal lattice it is not simply a dot or a spherical entity. It is a non-spherical particle of which the position *and orientation* are set up by the forces much more powerful then the forces of $H \leftrightarrow$ spin magnetic interaction. They are those of *chemical crystal bonding*. Orientations of the atomic particles cannot be affected by $H$ not only in the sense that its interaction with the spins is much weaker in comparison, but because changing of particle orientations cannot occur without a complete crystal rearrangement. Therefore, claiming that spontaneous magnetization *in the same crystal* can be redirected by the applied magnetic field actually means spin redirection *within* their atoms. Our assumption is that such a possibility does not exist. *Spins are uniquely oriented in the structure of their atoms.*

## 10. The Answers

The above-formulated assumption in conjunction with the outlined in Section 2 solid-state structural rearrangements clarifies and simplifies the whole problem of magnetization and, in fact, all other major problems of ferromagnetism. By treating that assumption as a fact, here are answers to the questions listed in Section 4.

■ **The cause of spontaneous magnetization**. The positions and orientations of the constituent particles of any crystal are those providing overall minimum free energy to it. Since spins are permanently oriented in the atoms, with the *particle* orientations come certain *spin* orientations. If all particles are packed in parallel, the crystal turns out to be spontaneously magnetized like that in Fig. 2. An antiparallel packing of spin-carrying particles will produce the antiparallel magnetic structure. A helical packing mode will produce the helical ferromagnet … This is how unsophisticated our explanation of the spontaneous magnetization and its diversity becomes.

■ **Thermodynamic stability of ferromagnets**. Magnetic spin interaction in ferromagnets increases their stability in some cases and decreases in others. But its contribution to the *total* crystal free energy dominated by the chemical bonding is relatively small. It is the *total* free energy that provides stability to a ferromagnet if it is the lowest among all other structural versions.

■ **Direction of spontaneous magnetization in crystal lattice. Physical nature of anisotropy energy.** It is not this mysterious energy, assumed to be specific to ferromagnetic crystals, that sets up direction of spontaneous magnetization. The direction is simply reflects the orientations of particles in the crystal. The "anisotropy energy" does not need explanation of its physical nature. It is not real.

■ **Magnetization by motion of domain boundaries.** Orientations of spins are fixed in their particles, and the orientations of particles are fixed in the crystal lattice. Therefore, spin orientations are not changeable without reconstruction of the crystal itself. Crystal reconstruction in solid state can proceed only by atomic/molecular rearrangement at interfaces. Magnetization (changing spin directions) by motion of domain interfaces is the only way to materialize. Magnetization "by rotation" is not possible.

■ **On "cooperative" spin rotation at the second stage of magnetization.** No spin rotation at whatever stage is possible.

■ **Domain formation and structure.** For the first time formation of a domain structure in its specific details could be explained. It is done in Section 7.

■ **Bloch wall.** Over and above all other arguments to reject the Bloch wall model, the following two reasons are sufficient to discard it: (1) The model was based on the erroneous idea of spin rotation in a crystal medium, and (2) The model was a derivative of two defunct theories: "exchange field" and "anisotropy energy".

■ **The real domain interface**. Everything said below is equally applicable to both 180°-boundary and any other domain interface. The domain boundary is a *structural twin interface* of zero thickness. The dipoles on the opposite sides of a 180°-domain interface are either antiparallel, or their projections on the interface are antiparallel. The experimental domain patterns are entirely consistent with the crystal twin nature of the domain interfaces. The energy of a twin interface, especially with the magnetic attraction, is minimal among any imaginable crystal interfaces. The twin interfaces propagate according to the *stepwise contact mechanism,* which is universal for all crystal rearrangements. This mechanism is the same whatever angle happened to be between the magnetic moments of the contacting domains. Relocation of one spin carrier takes place at a "kink" on the interface, one particle at a time. Every relocation results in reorientation of one magnetic moment. Such is the universal magnetization mechanism of ferromagnets.

■ **Magnetization in phase transitions.** Ferromagnetic phase transitions should be interpreted as a magnetization brought about by changing of the crystal structure (not vice versa). Over all 20th century they were classified as (and became the last resort for) the cooperative "second order" phase transitions [1, 4]. When accumulation of the experimental data has revealed that many of them were "accompanied by" structural rearrangements, the new term "magnetostructural" has come into use. The cause of that interconnection remained mysterious.

If the orientations of a spin and its carrier are uniquely bound, as argued in Section 9, the explanation becomes trivial: magnetization results from reorientation of the particles − spin carriers − upon the phase transition.

Vice versa: magnetization without reorientation of the magnetic particles is not possible. *All ferromagnetic phase transitions are structural* ("first order"), whether that has been detected of not yet. The term "magnetostructural" is superfluous.

This topic was considered in more detail in [4].

# 11. Magnetization of Antiferromagnet

Observations of magnetization of an antiferromagnet by Lavrov *at al.* (LKA) [29] illustrate the inability of standard science to account for them, while being in accord with our concept of magnetization.

LKA applied magnetic field to antiferromagnetic crystal and observed its rearrangement by motion of phase boundaries and kinks moving along them. While the kinks height, unlike those seen by Novoselov *et al.*, was not elementary, the magnetization was realized by the general molecular mechanism of solid-state reactions.

In terms of the conventional science the phenomenon itself had not existed. As LKA pointed out, "the common perception [is that] magnetic field affects the orientation of spins, but has little impact on the crystal structure." But the structure did changed. The perception is erroneous: the applied magnetic field was the driving force of phase transition. While not being able to turn spins directly in the same structure, its interaction with the spins was sufficient to upset the close balance of free energies between antiferro- and ferromagnetic phases. LKA observed a structural phase transition where change of magnetization was a secondary effect.

LKA noted that "one would least expect any structural change to be induced in antiferromagnet where spins are antiparallel and give no net moment". Nevertheless, the

phenomenon has took place. From the positions of the standard theory there are two reasons to see it as a contradiction. One is rooted in the belief that spins in an antiferromagnet are strongly bound together by the Heisenberg's "exchange forces", therefore the external field *H*, which is weak in comparison, cannot deal separately with the parallel and antiparallel components of the spin system. The other reason is a habitual approach to phenomena as being "cooperative" even when they are clearly not. LKA dealt with *antiferromagnet → ferromagnet* phase transition in which the particles fill "kinks" in sequence, every second one turning its spin toward the *H* direction.

# 12. Magnetic Hysteresis Loops

The inseparable connection of a magnetization process with the underlying structural rearrangements is a guiding principle for analyzing magnetic hysteresis loops. They, in fact, are *structural* hysteresis loops like those observed in the temperature solid-state phase transitions and other solid state reactions. Magnetic hysteresis loops are to be interpreted as the *structural rearrangements* driven by application of magnetic field, while the magnetization of the matter is recorded. The structural hysteresis loops, in turn, have their roots in nucleation and its hysteresis.

This scope of the interconnected phenomena cannot be argued meaningfully in the limited frames of the present article, but the topic is presented in detail in [1, 6].

# 13. Conclusions

It is shown that current theories of a spontaneously magnetized state and the magnetization process of ferromagnets are invalid. Their main components − Heisenberg's electron exchange field, anisotropy energy and Bloch wall, aside from their other inconsistencies, involve an implicit idea that spins (elementary magnetic dipoles) can be directly rotated by applied magnetic field, as if they are a separate free system. That is not the case. Importance of the crystal, which the spins belong to, was overlooked.

It is demonstrated that no sophisticated theory was needed. All problems of the theory are solved at once as soon as we accept the following: *the directions of spins are not changeable in the structure of atoms carrying them.* Then it becomes self-evident that the spontaneous magnetization is established by the *crystal field*, because it holds atoms and molecules in the crystal in certain positions *and orientations.* The only way to reorient spins is to reorient their atoms, and the only way to reorient atoms in solid state is atom-by-atom relocation at interfaces. No spin rotation in the same structure is possible. And so on.

Any change in the state of magnetization is simply an accompanying effect of the restructuring driven by application of magnetic field. The theories of electron exchange field, anisotropy energy and Bloch wall are both invalid and superfluous.

---

# REFERENCES

[1]   Y. Mnyukh, Fundamentals of Solid-State Phase Transitions, Ferromagnetism and Ferroelectricity, Authorhouse (2001) [or 2$^{nd}$ (2010) Edition].

[2]   Y. Mnyukh, arxiv.org/abs/1101.1249.

[3]   Y. Mnyukh, arxiv.org/abs/1103.2194.

[4]   Y. Mnyukh, Am. J. Cond. Mat. Phys. 2012, 2(5): 109-115. arxive.org/abs/1106.3795.

[5]   Y. Mnyukh, Am. J. Cond. Mat. Phys. 2013, 3(4): 89-103. arxiv.org/abs/1110.1654.

[6]   Y. Mnyukh, Am. J. Cond. Mat. Phys. 2013, 3(5): 142-150

[7]   C. Kittel, Ch. 16 *in* Introduction to Solid State Physics, 4$^{th}$ Ed., Wiley @ Sons (1971).

[8]   R.M. Bozorth, Ferromagnetism, Van Nostrand (1951).

[9]   A.I. Kitaigorodskii, Y. Mnyukh, Y. Asadov, J. Phys. Chem. Solids 26 (1965) 463.

[10]  Y. Mnyukh, N.N. Petropavlov, A.I. Kitaigorodskii, Soviet Physics - Doclady **11** (1966) 4.

[11]  Y. Mnyukh, N.I. Musaev, A.I. Kitaigorodskii, Soviet Physics - Doclady. 12 (1967) 409.

[12]  Y. Mnyukh, N.I. Musaev, Soviet Physics - Doclady 13. (1969) 630.

[13]  Y. Mnyukh, Soviet Physics - Doclady 16 (1972) 977.

[14]  Y. Mnyukh, N.N. Petropavlov, J. Phys. Chem. Solids 33 (1972) 2079.

[15]  Y. Mnyukh, N.A. Panfilova, J. Phys. Chem. Solids 34 (1973) 159.

[16]  Y. Mnyukh, N.A. Panfilova, Soviet Physics - Doclady 20 (1975) 344.

[17]  Y. Mnyukh et al., J. Phys. Chem. Solids 36 (1975) 127.

[18]  Y. Mnyukh, J. Crystal Growth 32 (1976) 371.

[19]  Y. Mnyukh, Mol. Cryst. Liq. Cryst. 52 (1979) 163.

[20]  Y. Mnyukh, Mol. Cryst. Liq. Cryst. 52 (1979) 201.

[21]  Y. Mnyukh, Am. J. Cond. Mat. Phys. 2013, 3(2): 25-30.

[22]  Y. Mnyukh, Am. J. Cond. Mat. Phys. 2014, 4(1): 1-12.

[23]  R.P. Feynman, R.B. Leighton, M. Sands, The Feynman Lectures on Physics v 2, Addison-Wesley, NewYork (1964).

[24]  Appendix in Ref [1].

[25]  J.H. Samson, Phys. Rev. B 51 (1995) 223.

[26]  A. Hubert, Theorie der Domanenwande in Geordneten Medien, Springer-Verlag (1974).

[27]  Y. Mnyukh, Am. J. Cond. Mat. Phys. 2014, 4(3): 57-62; arxiv.org/abs/1103.4527.

[28]  K.S. Novoselov, A.K. Geim, S.V. Dubonos, E.W. Hill and I.V. Grigorieva, Nature 426, (2003) 812.

[29]  A.N. Lavrov, Seiki Komiya and Yoichi Ando, Nature 418 (2002) 385; (http://arxiv.org/abs/cond-mat/0208013.

# Paramagnetic State and Phase Transitions

This article is complementary to the article "Ferromagnetic State and Phase Transitions" [1]; it is about paramagnetism in solids. Magnetic elements of the periodic system are currently classified into ferromagnetic and paramagnetic types. All of them, however, can exist in one or the other crystal phase depending on the thermodynamic conditions. They actually constitute a single − magnetic − type by possessing of a spin. Evidence is presented that random orientation of spins in paramagnetic crystals results from thermal rotation of the spin-carrying particles (atoms, molecules), and not spins themselves. Connection ofa paramagnetic crystal state to the phenomenon oforientation-disordered crystals is revealed and its significance for the theory of paramagnetism is emphasized.

American Journal of Condensed Matter Physics 2015, 5(2): 56-59
DOI: 10.5923/j.ajcmp.20150502.03

# Paramagnetic State and Phase Transitions

## Yuri Mnyukh

76 Peggy Lane, Farmington, USA

**Abstract**   This article is complementary to the article "Ferromagnetic State and Phase Transitions" [1]; it is about paramagnetism in solids. Magnetic elements of the periodic system are currently classified into *ferromagnetic* and *paramagnetic* types. All of them, however, can exist in one or the other crystal phase depending on the thermodynamic conditions. They actually constitute a single − magnetic − type by possessing of a spin. Evidence is presented that random orientation of spins in paramagnetic crystals results from *thermal rotation of the spin-carrying particles* (atoms, molecules), and not spins themselves. Connection of a paramagnetic crystal state to the phenomenon of *orientation-disordered crystals* is revealed and its significance for the theory of paramagnetism is emphasized.

**Keywords**   Magnetic type, Paramagnetic, Ferromagnetic, Phase transitions, Orientation-disordered crystals

## 1. Introduction

The remarkably well designed tables [2] by Wolfram Research, Inc. (WRI) on the periodic system of chemical elements sort out the elements by magnetic type. There 32 elements are classified as *paramagnetic type*, while Fe, Co, Ni and Gd as *ferromagnetic type* to which Cr − the only element marked "antiferromagnetic" − should be added as a particular case of ferromagnetism. It may appear that being of one or the other type is rooted in the elements themselves. Whether it is the case will now be considered with an emphasis on the physics of a crystal paramagnetic state.

## 2. Paramagnetic State of Ferromagnetic Elements

The elements of the two magnetic types feature an elementary magnetic moment ("spin") − against the remaining members of the periodic system. No other magnetic difference between the *elements* of that combined group is directly envisioned. Still, if it is real, it could be only between the crystals they form. At this point the phenomenon is to be invoked that ferromagnetic crystals of Fe, Co, Ni, Gd and Cr turn paramagnetic at elevated temperatures (called "Curie points"). For example, it is ~770°C in Fe, ~1121°C in Co and ~ 358°C in Ni. Different

interpretation of that phenomenon stems from two disparate theories of ferromagnetic state and *ferromagnetic* ↔ *paramagnetic* phase transitions.

## 3. The Old Theory

According to the old theory (*e.g.*, [3, 4]), the spontaneous magnetization (regular alignment of spin directions) of ferromagnetic crystals results from spin interaction in the "electron exchange field" that supposedly existed due to overlapping of the electron shells. Ordering effect of the exchange field is opposed by thermal agitation and, at elevated temperatures ("Curie points", Table 1), the ferromagnetic spin order is destroyed. The resulting crystal phase is identified as *paramagnetic* by its ability to be magnetized in the direction of the applied field $H$, falling to zero at $H = 0$. This theory views the spin system as a free entity bound to neither the atoms, nor the crystal it belongs.

In contradiction with that view, the spin-carrying atoms emerge in the adjacent part of the theory dealings with a paramagnetic effect − predominance of orientations of atomic magnetic moments $M$ in the direction of applied magnetic field $H$. The magnetic atom is regarded there a *particle of a certain structure and direction of its magnetic moment $M$*. The applied field $H$ does not change the angle $M^{\wedge}H$, but makes the atomic magnetic axis (and, by implication, the atom) precessing around $H$ under the same angle. Because this does not produce a paramagnetic effect, atomic collisions in a gas medium had to be invoked. The theory has not placed magnetic atoms into the crystal lattice where they are in a close contact with the surrounding atoms.

* Corresponding author:
yurimnyukh@att.net (Yuri Mnyukh)
Published online at http://journal.sapub.org/ajcmp

**Table 1.** Data excerpted from the *Wolfram Research Inc.* tables "Magnetic Type", "Curie Point" and "Neel Point"

| Element | Atomic No. | Name | Magnetic Type | Curie Point (°K) | Neel Point (°K) |
|---------|-----------|------|---------------|------------------|-----------------|
| Cr | 24 | Chromium | Antiferromagnetic | | 393 |
| Mn | 25 | Manganese | Paramagnetic | | 100 |
| Fe | 26 | Iron | Ferromagnetic | 1043 | |
| Co | 27 | Cobalt | Ferromagnetic | 1394 | |
| Ni | 28 | Nickel | Ferromagnetic | 631 | |
| Ce | 58 | Cerium | Paramagnetic | | 125 |
| Nd | 60 | Neodymium | Paramagnetic | | 192 |
| Sm | 62 | Samarium | Paramagnetic | | 106 |
| Eu | 63 | Europium | Paramagnetic | | 90.5 |
| Gd | 64 | Gadolinium | Ferromagnetic | 292 | |
| Tb | 65 | Terbium | Paramagnetic | 222 | 230 |
| Dy | 66 | Dysprosium | Paramagnetic | 87 | 178 |
| Ho | 67 | Holmium | Paramagnetic | 20 | 132 |
| Er | 68 | Erbium | Paramagnetic | 32 | 82 |
| Tm | 69 | Thulium | Paramagnetic | 25 | 56 |

## 4. The New Theory

The new theory [1, 5] was put forward since it became evident that the standard theory was inadequate and contradictory. Thus, the old theory was only about parallel and antiparallel spin alignments, but was unable to explain the actual unlimited diversity of spin patterns ("magnetic structures"); its predictions vastly disagreed with experiment; magnetization by interface propagation was foreign to it; the "electron exchange field" produced a wrong sign of spin interaction; the "Curie points" revealed hysteresis… and much more.

The new theory maintains that ferromagnetic spin order is established by crystal forces. Indeed, it is the essence of a crystal matter to arrange both positions *and orientations* of its constituent particles in the long-range order according to minimum free energy of their chemical bonding. A simple fact is stressed that every spin belongs to its atomic carrier. Atoms of atomic crystal lattice function as particles of a certain geometrical shape. *The chemical bonding orients atomic spin carriers and in doing so it creates a ferromagnetic spin order.* Crystallographic orientation of the *spin carriers* (atoms, ions, molecules) in ferromagnetic crystals of any kind gives rise to the observed diversity of spin patterns (magnetic structures): parallel, antiparallel, ferr*i*magnetic, spiral, etc.; antiferromagnetism is not a phenomenon opposite to ferromagnetism, but a particular case of it.

Still, a ferromagnetic crystal state can only be explained in the frames of the new theory if *spins are invariably oriented in the structure of their atoms*. But there is no physical reason to doubt it is not so. As said above, it was used by the classical theory of paramagnetic effect without arguments. More importantly, it has led to the new comprehensive account both for the stability of a ferromagnetic state and for its all manifestations [1, 5]. Finally, the actual mechanism of the *ferromagnetic ↔ paramagnetic* transitions indicates whether spins alone, or their carriers are rotating in paramagnetic crystals. If the former, the crystal structure remains intact. If the latter, the crystal must be reconstructed. Experimentally, the latter is found to be true. The *ferromagnetic ↔ paramagnetic* transitions, including those in Fe, Ni and Co [5, 6], materialize by nucleation and growth, meaning that they are *structural* phase transitions. (According to the now obsolete, but still used terminology, they are "first order").

## 5. Paramagnetic Orientation-Disordered Crystals

The novel explanation of a ferromagnetic state throws different light on the physical nature of paramagnetic crystals: *it is the atomic spin carriers, and not spins themselves, are engaged in the thermal rotation.* Paramagnetic crystals are simply a case of the currently well-known phenomenon called *orientation-disordered crystals* (ODC), not related to their magnetic properties [7-9]. The constituent particles in the ODC are engaged in a thermal rotation, hindered to one or another degree − from jumping between a number of fixed orientation positions to almost smooth rotation. The phenomenon is observed in different kind materials and is more common than meets the eye. Phase transitions CRYSTAL → ODC materialize by the universal nucleation-and-growth mechanism of crystal rearrangements [5, 10] and accompanied by increase in the specific volume (2-3% in molecular crystals, less in other cases), allowing the closely packed system of particles to go into a rotating mode. Paramagnetic state of the compounds containing magnetic atoms could be due to: (a) rotation of the whole molecules, (b) rotation of the molecular branches containing those atoms, (c) rotation of only those atoms, (d) combination of some or all those rotations.

Representation of paramagnetic crystals as ODC raises question whether the classical theory of a paramagnetic effect is applicable. Considering that a close molecular packing in the ODC state is basically preserved, molecular rotation there is hindered. Within a short range it is reminiscent to interconnected rotating gears. Precession of atoms and molecules in that environment is hardly possible. On the other hand, existence of the orientational effect $M \rightarrow H$ under these conditions seems qualitatively self-evident. As for a quantitative theory, its development will be impeded by the multitude of the types and freedom degrees of the molecular rotation.

## 6. Ferromagnetic State of Paramagnetic Elements

After the foregoing clarification of physical nature of the paramagnetic phase produced by the crystals of "ferromagnetic-type" elements at high temperatures, the question about specifics of the 32 "paramagnetic-type" elements remains. To answer, it should be first noticed that five of them − rare earth elements (atomic No. 65-69) − exhibit Curie points as well, but located at low temperatures. However, our human perception of low and high temperatures should not be applied. Below 20 K all of them will join their counterpart Gd to become *ferromagnetic*. Here we deal again with the "low-temperature" ferromagnetic and "high-temperature" paramagnetic phases. Their low-temperature Curie points are high enough for particles of the paramagnetic phases to be engaged in thermal rotation. The five "paramagnetic" rare-earth elements do not constitute a separate magnetic type.

There are also 5 "paramagnetic" elements (atomic No. 25, 58, 60. 62, 63) with "Neel points". Physical meaning of these "points" is identical to the Curie points, namely, to indicate the temperature separating ferromagnetic and paramagnetic phases, the difference in the names being relates only to history of science. These 5 elements do not constitute a separate magnetic type either. Both "points", however, are identically incorrect: they are not *points*, but temperature *ranges* of structural phase transitions, a subject of hysteresis [5, 10-12].

According to the WRI tables, the rest 22 "paramagnetic" elements do not reveal ferromagnetic phase at any temperature. But that does not mean the ferromagnetic area is not present in their *temperature T − pressure p* phase diagrams. Considering that change to the rotation mode involves an increase in the specific volume, application of

sufficiently high pressure would transfer them to ferromagnetic phase.

## 7. Conclusions

1. The elements currently classified as either *ferromagnetic* or *paramagnetic* type are actually constitute a single magnetic type that differs from all other elements by possession of the elementary magnetic moment ("spin"). They are ferromagnetic or paramagnetic in different parts of the $T - p$ phase diagram. The classification shows only in which crystal phase they are found at room temperature and atmospheric pressure.

2. Magnetic properties of paramagnetic crystals result from random orientation of the *magnetic particles* (atoms and molecules) due to their thermal rotation in the crystal lattice, and not rotation of spins alone as usually implied.

3. Paramagnetic crystals are a case of the *orientation-disorder crystals* (ODC) not specifically related to the magnetic properties. The classical theoretical considerations about precession of magnetic atoms in a gas phase when magnetic field is applied are not applicable to the paramagnetic crystals.

4. *Ferromagnetic ↔ paramagnetic* phase transitions materialize by the nucleation-and-growth mechanism general to all solid-state phase transitions. They are not of a "critical" ("second order" [10]) nature.

5. It is predicted that the "paramagnetic-type elements" will produce ferromagnetic phase if sufficiently high pressure applied to their crystals.

---

## REFERENCES

[1] Y. Mnyukh, "Ferromagnetic state and phase transitions", Am. J. Cond. Mat. Phys. 2012, 2(5): 109-115; arxive.org/abs/110 6.3795.

[2] http://periodictable.com/Properties/A/MagneticType.html.

[3] C. Kittel, Introduction to Solid State Physics, 4th Ed., Wiley, (1971).

[4] I.E. Tamm, Fundamentals of the Theory of Electricity, Mir Publications, Moscow (1979).

[5] Y. Mnyukh, Fundamentals of Solid-State Phase Transitions, Ferromagnetism and Ferroelectricity, Authorhouse, 2001 [or 2nd (2010) Edition].

[6] Sen Yang *et al.*, "Evidence for first-order nature of the ferromagnetic transition in Ni, Fe, Co and CoFe$_2$O$_4$", Phys. Rev. B 8, 174427 (2008).

[7] N.G. Parsonage and L.A.K. Staveley, Disorder in Crystals, Clarendon Press (1978).

[8] N. Sherwood, Plastically Crystalline State: Orientationally Disordered Crystals (1979).

[9] Y. Mnyukh, "Order-disorder phase transitions as crystal growth", Section 2.7 in Ref. [5].

[10] Y. Mnyukh, "Mechanism and kinetics of phase transitions and other reactions in solids", Am. J. Cond. Mat. Phys. 2013, 3(4): 89-103; arxiv.org/abs/1110.1654.

[11] Y. Mnyukh, "Hysteresis and nucleation in condensed matter", arxiv.org/abs/1103.2194.

[12] Y. Mnyukh, "Second-order phase transitions, L. Landau and his successors", Am. J. Cond. Mat. Phys. 2013, 3(2): 25-30; arxiv.org/abs/1102.1085.

# The True Cause of Magnetostriction

The cause of magnetostriction is revealed by analyzing this phenomenon in a polydomain crystal of Fe. It is based on the two fundamentals: (a) magnetization is a rearrangement of spin directions by rearrangement of the crystal structure, and (b) the $\alpha$-Fe has a tetragonal rather than cubic structure. The magnetostriction results from bringing the shorter tetragonal axis to coincide with, or closer to the direction of the applied magnetic field. It is not rooted in the alleged deformation of crystal unit cell.

American Journal of Condensed Matter Physics 2014, 4(3): 57-62
DOI: 10.5923/j.ajcmp.20140403.03

# The True Cause of Magnetostriction

## Yuri Mnyukh

76 Peggy Lane, Farmington, CT, USA

**Abstract**    The cause of magnetostriction is revealed by analyzing this phenomenon in a polydomain crystal of Fe. It is based on the two fundamentals: (a) magnetization is a rearrangement of spin directions by rearrangement of the crystal structure, and (b) the α-Fe has a tetragonal rather than cubic structure. The magnetostriction results from bringing the shorter tetragonal axis to coincide with, or closer to the direction of the applied magnetic field. It is not rooted in the alleged deformation of crystal unit cell.

**Keywords**    Magnetostriction, Ferromagnetism, Magnetization, Iron, Tetragonal, Domain structure, Deformation

## 1. Introduction

Magnetostriction was first observed by J. Joule in 1842. Currently it is typically described as "deformation of a body" as a result of its magnetization. The elongation or contraction in the direction of applied magnetic field $\pm (\ell-\ell_o) / \ell_o$ is usually between $10^{-5}$ and $10^{-3}$ and accompanied by the opposite sine changing in the transverse direction, so that the volume remains almost the same. The above definition is not incorrect, but we would prefer not to use the word "deformation", since it unwittingly suggests (erroneously as will be shown) to ascribe the phenomenon to deformation (also called "distortion") of its crystal lattice. We define the magnetostriction simply as change of shape of a ferromagnetic body by applied magnetic field.

Magnetostriction involves no change of crystal structure and not to be confused with the changes taking place in phase transitions, especially those driven by application of magnetic field. No credible theory of the phenomenon exists. Here is how the current status of the problem is described in [1]: "Despite the tremendous advances in modern electronic structure theory for studies in material science, magnetostriction has been rarely attacked until recently due to its intrinsic complexity". It will now be shown that the seeming complexity of the phenomenon results from looking for the answer in wrong direction. The secret of its origin is hidden in the *magnetization process*, provided it is properly understood.

The ample literature on magnetostriction is devoted to its observation in different materials, its technological applications, and its negative effects. Attempts of its explanation are rare, superficially-descriptive, brief and vague, but unanimous in the belief that magnetostriction results from elastic deformation / distortion of the crystal lattice strained by forces of magnetic interaction. To illustrate these views we turn to two renowned authorities in ferromagnetism.

K. P. Belov [2]: "*Magnetization, which occurs by displacement of domains boundaries and rotation of domain magnetic moments, leads to changing of equilibrium distances between the atoms of the lattice; the atoms are shifted and magnetostrictive deformation of the lattice occurs*".

This explanation is erroneous in every its point (see Sections 2 and 3). It incorrectly describes the mechanism of magnetization. It is not specified why interface propagation will change the inter-atomic distances (it will not). Rotation (reorientation) of magnetic moments in a domain is not possible. Deformation of the lattice was merely a guess: it does not occur.

S. Chikazumi [3]: "The reason for [magnetostriction] is that the crystal lattice inside each domain is spontaneously deformed in the direction of domain magnetization and its strain axis rotates with the rotation of the domain magnetization, thus resulting in a deformation of the specimen as a whole".

Noticeably, here magnetization by motion of domain boundaries − an undeniable experimental fact − is not even mentioned. A further commenting on this excerpt would be a repetition.

*In both cases (and everywhere else) the crystal lattice is assumed to become strained. The idea that magnetostriction is due to elastic deformation of crystal lattice is so entrenched that it even entered into this definition: [magnetostriction is] "dependence of the state of strain (dimensions) of a ferromagnetic sample on the direction and extent of its magnetization*" [4]. No attention is being paid to the fact that removing the magnetic field does not lead to elastic relaxation of the sample towards its "unstrained" condition. Why? It is simply not strained.

## 2. Fundamentals of Ferromagnetism: New *vs.* Old

Magnetostriction cannot be correctly explained until the physics of magnetization is understood. The currently dominated theory of ferromagnetism, based on the Heisenberg's idea on extremely strong *electron exchange interaction*, has failed to account for thermodynamic stability of ferromagnetic state and for basic ferromagnetic phenomena [5, 6]. As for the magnetization is concerned, the current theory involves a belief that it is realized by a "rotation" (meaning: reorientation) of spins *in* the crystal structure. The theory is powerless to explain why magnetization proceeds by propagation of domain boundaries.

The new physics of ferromagnetism was put forward in 2001 [5] and formulated again in [6]. Here are its main principles:

### 1. Stability of ferromagnetic state.

The Heisenberg's extremely strong "field of electron exchange forces" in a ferromagnetic materials does not exist. Only the classical magnetic interaction is real; its contribution to the crystal total free energy is relatively small as against the chemical crystal bonding (and by itself has usually a destabilizing effect). A simple consideration that the *total* free energy of a ferromagnetic crystal determines its stability eliminates the central problem the previous theory was unable to solve: a *ferromagnetic crystal is stable due to its low total free energy in spite of a possible destabilizing effect of the spin magnetic interaction.*

### 2. Orientation of spins in their carriers.

*Orientation of a spin is uniquely bound to the orientation of its atomic carrier.* The spatial distribution and orientation of spins (*i.e.* state of magnetization) is *imposed* by the crystal structure (by its chemical bonding). It also follows that any change of spin orientation in a crystal (in other words, any magnetization) requires reorientation of the carriers. That can occur only by restructuring of the crystal itself. The general mechanism of a crystal restructuring in solids is nucleation and molecule-by-molecule rearrangement at interfaces [5, 7]. That is why any magnetization proceeds only by interface propagation, while spin "rotation" (reorientation) in the otherwise intact crystal lattice is impossible [5, 6, 8]. On the same reason all ferromagnetic phase transitions (i.e. those involving magnetization) are *structural*. They are "first order" rather than "second order" [5, 9].

### 3. Essence of paramagnetic state.

Paramagnetic state is the orientation-disordered crystal (ODC) state known in other types of crystals, where a translation 3-D crystal order exists, but atoms or molecules − spin carriers − are engaged in a thermal rotation resulting in

zero spontaneous magnetization.

## 3. Magnetization of Iron

The physical origin of magnetostriction will be demonstrated by showing how it emerges in iron. The phenomenon easily reveals itself in a common technical Fe. It is known that after melt solidification and cooling down to room temperature, the common solid Fe consists of arbitrarily shaped and oriented grains. Every grain is not a single crystal, but a *polycrystal* − a complex of peculiarly organized single-crystal domains separated by straight boundaries. It is a structural reorganization of the domain complex, and not a deformation of the Fe unit cell, will be shown to give rise to the magnetostriction.

Iron is a ferromagnetic material, but, as known, can be either in a ferromagnetic (F) or non-ferromagnetic (paramagnetic, P) state. Depending on the state, it reacts differently to magnetic field $H$. In the F-state the resultant magnetization ($M$) will be either $M \neq 0$, or changed from $M_1$ to $M_2$; in the P-state $M \equiv 0$ under all circumstances.

In terms of the principles listed in Section 2 we are able to demonstrate the structural mechanism of magnetization by using two neighboring domains by way of example (Fig. 1). They are single crystals of identical crystal structure, naturally magnetized to saturation, but differently oriented together with their magnetization vectors $M_S$. Application of magnetic field $H$ to this system cannot turn $M_S$ toward $H$ directly, considering that $M_S$ is fixed by crystal structure. It upsets the balance of the domain free energies $E$, initially identical. The domains react as if they are different crystal phases: the one of a lower free energy (No. 1) becomes preferable and grows by a movement of the interface AB to right into No. 2. *Magnetization of the system is realized without a "rotation" of the whole domains* (sometimes suggested, but unimaginable*) or reorientation of spins in the otherwise intact crystal structure* (which is impossible − as stated in Section 2).

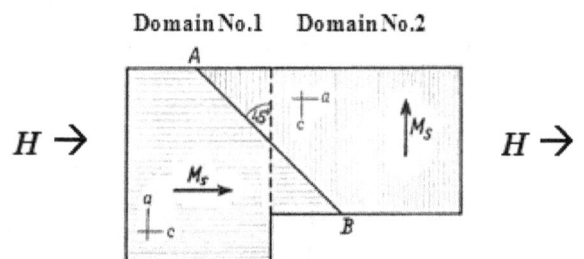

**Figure 1.** Magnetization process in Fe by propagation of interface AB between the two spontaneously magnetized domains of identical crystal structure. The AB, called "90°- border" is a crystallographic twin of zero thickness where the structures perfectly match. Application of magnetic field $H$ makes the free energy of No.1 lower than No. 2, causing it grow by AB movement to right, consuming No. 2. However, magnetization of a solitary crystal needs nucleation

The same approach accounts for $M \equiv 0$ when material is in its P-state. Application of magnetic field does not upset the energy balance $E_1 = E_2$ due to thermal rotation of *atoms* and their spins in both domains, so the interface remains still. However, if spins could freely change their orientations because not being bound to the orientations of their atomic carriers, at least some orientation effect (magnetization) toward **H** due to direct magnetic interaction could be expected.

Concluding, it is to be noted that

(a) the described magnetization process is based on the same principle of crystal growth as that taking place in solid-state phase transitions, even though here no change in the type of crystal structure is involved, and

(b) the cause of magnetization by interface propagation gets its natural explanation.

## 4. The Crystal Structure of Fe

The crystal structure of the room-temperature α-phase of Fe will be reexamined. Over the whole 20th century the phase behavior of Fe (Fig. 2, left side) looked enigmatic [10]. There were three different crystal phases α, γ and δ on the temperature scale. The room-temperature α-phase, believed to be truly cubic, bcc, $a = 2.86645$ Å at 20℃ [11], stretched up the temperature scale to 910℃ where it undergoes phase transition to γ-phase [12]. The α-phase above ~ 770℃ is paramagnetic, but ferromagnetic below it. The paramagnetic part of the α(bcc) phase is called "β-phase" only conditionally and could not be clearly categorized [13-16]: "Beta ferrite (β-Fe) and beta iron (β-iron) are obsolete terms for the paramagnetic form of ferrite (α-Fe)"..."Beta ferrite is crystallographically identical to alpha ferrite, except for magnetic domains"..."The beta 'phase' is not usually considered a distinct phase but merely the high-temperature end of the alpha phase field".

Others call the ferromagnetic α↔β change a "second order phase transition". It was in violation of what the second-order phase transitions assumed to be, because "transition from the ferromagnetic to the paramagnetic state is a phase transition of a second kind. At the Curie point, where the spontaneous magnetization disappears, the symmetry of a ferromagnet changes sharply" [17]. If so, can the ferromagnetic α(bcc) → α(bcc) transition be "second order"?

In terms of the conventional theory of ferromagnetism this picture was enigmatic indeed. It led Myers [10] to state: "Unusual and nontypical, elemental iron can provide the impetus for discussing ... basic thermodynamic concepts and the phenomenon and theory of ferromagnetism".

The required clarification has come from the new concepts of ferromagnetism referred in Section 2. All phases on the temperature scale (Fig. 2), including the β-phase, are crystallographically different and transit to their neighbors by nucleation and rearrangements at interfaces. The α-phase occupies the temperature range only from ~769℃ down. All the phase transitions are first-order, exhibiting phase

coexistence, hysteresis and latent heat − detected or not yet. The phases β, γ and δ are ODCs. They are paramagnetic due to thermal rotation of the atomic spin carriers.

The existence of the β → γ and γ → δ, phase transitions ODC → ODC above ~769℃ is an indication that the atomic spin carriers are not spheroid-shaped entities and their hindered rotation required more space with increasing temperature. On that background the CRYSTAL → ODC transition α → β occurring without changes in density, latent heat, and other characteristics of first-order phase transitions, if true, would be an anomaly. But the above characteristics of this transition were incorrect. As has been concluded in [5], the room-temperature α-phase emerges upon cooling from the β-phase by nucleation and growth of a new crystal structure. While this change was not noticed for a very long time, the latent heat − signature of a structural transition − was ultimately recorded [18].

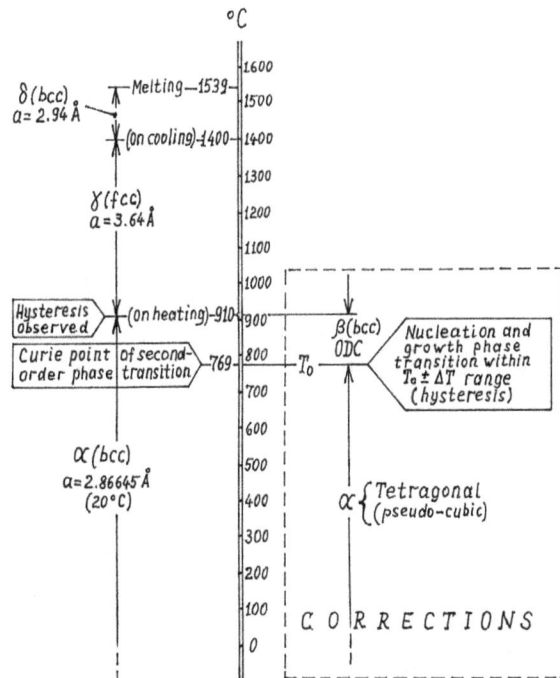

**Figure 2.** Phase transitions in Fe as they believed to be. On the right - corrections eliminating existing contradictions. The difference is in the interpretation of the α-structure and α - β ferromagnetic transition at ~769℃. ODC is paramagnetic orientation-disordered crystal state

We shall take a close look at the crystal lattice of α-Fe within a domain. Its unit cell and the directions of spontaneous magnetization are shown in Fig. 3. *The spin magnetic interaction is an integral part of the forces establishing equilibrium interatomic distances. The structure is not strained. It is in its natural stable state.* While its symmetry was deemed to be cubic on the basis of precision measurements of the unit cell parameters, it cannot be so according to the rules of symmetry. Indeed, the uniform orientation of spins results from the uniform orientation of their atomic carriers located in the sites of the unit cell. Such crystal lattice lacks of cubic centre of symmetry, leaving it to be a pseudo-cubic. The unit cell is not truly cubic geometrically as well. The mutual repulsion of the parallel elementary dipoles in the two equivalent orthogonal directions $a$ and $b$ of the pseudo-cubic unit cell makes its $c$-parameter in the third orthogonal direction the shortest one, thus producing a tetragonal unit cell with $c < a = b$. The difference is minor, but it is sufficient to produce the observed magnetostriction.

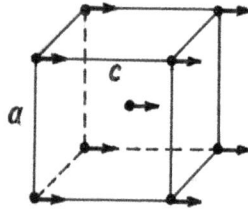

**Figure 3.** Unit cell of ferromagnetic α-Fe crystal and its spontaneous magnetization. It is tetragonal with $c < a = b$. The difference is very small, but enough to be the source of the observed magnetostriction

## 5. The Domain Structure of Fe

Every grain of technical Fe is a polydomain crystal comprised of single-crystal domains arranged in a specific pattern of a kind shown in Fig. 4. They are divided by straight twin interfaces. Applied magnetic field causes the interfaces to move in coordination. In three dimensions, the spins of every domain belong to one of six mutually orthogonal directions (Fig 5). At that, the spins of two neighboring domains form either 90° ("90° boundary") or 180° ("180° boundary").

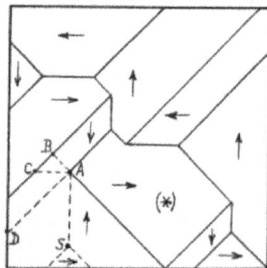

**Figure 4.** A schematic illustrating the features of domain patterns in the Fe polydomain crystals (shown is (*100*) plane). The broken lines help to comprehend their "rules": only interface AS is correct ([5], Sec. 4.9)

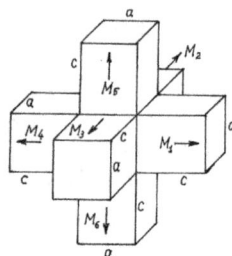

**Figure 5.** Six mutual directions of spontaneous magnetization $M_S$ that make 90° and 180° to one another in a polydomain iron crystal. While the unit cell dimensions $c$ and $a = b$ are very close, it is axis $c$ which represents the $M_S$ direction. (The figure is not intended to illustrate the orientations of domain boundaries)

In a polydomain Fe crystal the $c$-axes of the domains are initially distributed over the six equivalent spatial directions. In the applied magnetic field the domain structure undergoes rearrangements by growth of the "better oriented" domains at the expense of others. In average, the domain structure acquires preference of its $c$-axis in the direction closer to the applied field $H$, somewhere between 0° and 45° to $H$, depending on the $H$ strength and other conditions.

## 6. Magnetostriction is Illustrated

Misinterpretation of magnetization, as described above, has been a major barrier to the identification of the cause of magnetostriction. Another obstacle, now also removed, was not detecting the unit cell of α-Fe to be a tetragonal rather than cubic.

The cause of the magnetostriction in a Fe polydomain crystal is illustrated in Fig. 6. Bozorth [19] indicated that a "magnetization reversal" does not produce magnetostriction, but changing of the spin direction by 90° does. The part A, B of Fig. 6 explains that. The $c$-axes of the domains in the case of 180° boundary remain aligned along $H$ both before and after the crystal rearrangement. No geometrical change is produced. But the structural rearrangement resulting in alignment of the $c$-axes along $H$ in case of the 90° boundary gives rise to changing of the body length.

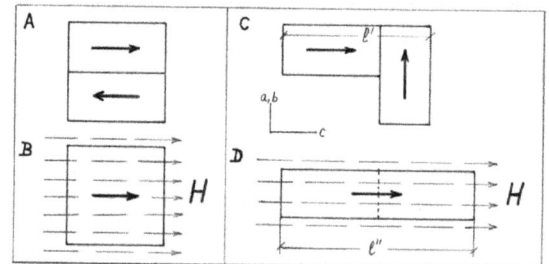

**Figure 6.** Sketch showing a system of two domains of the shape reflecting their tetragonal unit cell. For the sake of illustration the parameter $c$ is made much longer than $a = b$. (A,B): Magnetostriction of the antiparallel domains is zero, as actually found experimentally. (C,D): In case of the domains making initially 90° with each other, rebuilding the domain on the right into a position with $c \parallel H$ gives rise to the magnetostriction $\ell'' - \ell'$

## 7. Why Tetragonal Fe Has Not Been Detected

Quantitatively, difference between the $c$ and $a$ parameters was not large enough to be directly detected. But it was sufficient to be seen and measured as a cumulative effect called magnetostriction. Inasmuch as this phenomenon is a change in the shape of the same quantity of the matter, the length magnetostriction is accompanied by a transverse magnetostriction of the opposite sign, leaving very little to the *volume* magnetostriction. The latter is, evidently, a secondary effect accompanying any structural rearrangement.

We can roughly estimate the difference between the $c$ and $a$ parameters of an α-Fe unit cell by ascribing it to its reorientation by 90°. The maximum length magnetostriction (defined as fractional elongation or contraction) in Fe, reported in a number of experimental studies, including performed on single crystals, was about $2 \cdot 10^{-5}$, which gives us

$$(a - c) = 2.86645 \text{ Å} \cdot 2 \cdot 10^{-5} \approx 0.00006 \text{ Å}.$$

The value $a = 2.86645$ Å of a Fe unit cell, supposed to be cubic, was the result of special precision measurements in 1948 [11] by the best technique available at that time. It was a powder X-ray photography by recording back reflections on film in circular camera. The reported probable error of a single observation was ±0.00003Å with Co radiation and ±0.00006 Å with Cr radiation. Thus our estimate for $(a - c) \approx$ 0.00006 Å, even if it is several times greater in reality, counts at the very limits of the precision given for $a = 2.86645$ Å. More importantly, the reflections were indexed in a cubic lattice, so an excessive widening of the reflection line (and those illustrated in the work are far ftom being sharp) was not taken into consideration. At that, the center of this line, being, evidently, a superposition of two partially overlappind lines, could be measured with high precision. A a result, the deviation of the a-Fe unit cell from a geometrically cubic could easily be overlooked.

## 8. "Electrostriction" is Differently Defined

The counterpart of magnetostriction in ferroelectrics had to be "*electrostriction*", but this is not the case. The latter notion was defined differently – not specific to ferroelectrics. As opposed to the definition of magnetostriction, this name includes a small elastic deformation of any dielectrics when they are placed in an electric field $E$. Because the (+) and (-) charges of the electric dipoles are, or can be, spatially separated in the crystal unit cell of a dielectric, application of $E$ either causes its polarization by crystal rearrangement, or affects the value of its dipole moments. The latter produces some elastic deformation of the crystal. The analogous effect - induced magnetization and the corresponding elastic deformation - is essentially absent in ferromagnetics where the poles of the elementary magnetic moments cannot be split apart.

Even though the theory does not have a name for the ferroelectric counterpart of magnetostriction, it is real. Let us name it *ferroelectrostriction* (FES). The essence of what was said about magnetostriction holds true for FES. The case of $BaTiO_3$, for instance, is rather similar to that of Fe. The cubic paraelectric phase of $BaTiO_3$ becomes tetragonal below 120℃ after its (nucleation and growth) phase transition into the ferroelectric phase. The only difference is that the $c$-parameter differs from the $a = b$ by 1%, *i.e.*, much greater than in Fe. Formation of the domain structure by nucleation in the six fixed orientations within the cubic paraelectric

matrix occurs in the same manner. Domain rearrangements during polarization in electric field give rise to FES. The FES is a change in the shape of a polydomain ferroelectric body as a result of crystal rearrangements of the domains in the process of polarization under the action of applied electric field. In the combined effect of the induced polarization and FES, the latter is greater.

## 9. Conclusions

The simple crystal structure of Fe and the simple pattern of its spontaneous magnetization were very helpful in analyzing and illustrating the cause of magnetostriction.

The previous views on this phenomenon as deformation / distortion of crystal structure by magnetic field were based on misinterpretation of a magnetization process as reorientation of spins in the crystal lattice. On this reason a part of the present article was devoted to that important aspect of ferromagnetism. It was illustrated that magnetization is actually realized by a domain reconstruction at their interfaces according to nucleation-growth principle.

Another critical component in our explanation of the magnetostriction was to reveal a tetragonal rather than cubic crystal structure of Fe. It is the restructuring of the domain system toward predominant alignment of the tetragonal axes $c$ in the direction of the applied magnetic field that gives rise to the change of the body dimension. No intrinsic deformation / distortion of crystal lattice involved in that process.

## REFERENCES

[1]   R. Wu, V.I. Gavrilenko and A.J. Freeman, in *Modern Trends in Magnetostriction Study and Applications,* Ed. M.R.J. Gibbs, Kluwer Acad. Press (2000).

[2]   K. P. Belov, *Magnetostriction,* in *The Great Soviet Encyclopedia, 3rd Edition (1979).*

[3]   *S. Chikazumi,* Physics of Ferromagnetism, *Oxford Univ. Press (2009).*

[4]   *McGraw-Hill Dictionary of Scientific & Technical Terms,* 6th Ed. (2003).

[5]   Y. Mnyukh, *Fundamentals of Solid-State Phase Transitions, Ferromagnetism and Ferroelectricity,* 2001 [or 2nd (2010) Edition].

[6]   Y. Mnyukh, *Amer. J. Cond. Mat. Phys.* 2012, 2(5): 109.

[7]   Y. Mnyukh, *Amer. J. Cond. Mat. Phys.* 2013, 3(4): 89.

[8]   Y. Mnyukh, arxiv.org/abs/1101.1249.

[9]   Y. Mnyukh, *Amer. J. Cond. Mat. Phys.* 2013, 3(2): 25.

[10]  C. E. Myers, *J. Chem. Educ.* 43, 303–306 (1966).

[11]  D E Thomas, *J. Sci. Instrum.* 25 (1948) 440.

[12]  J. Donohue, *The Structure of Elements*, John Wiley & Sons (1974).

[13]  D. K. Bullens *et al.*, *Steel and Its Heat Treatment*, Vol. I, 4th Ed., J. Wiley & Sons Inc., 1938.

[14]  S. H. Avner, *Introduction to Physical Metallurgy,* 2nd Ed., McGraw-Hill, 1974.

[15]  *ASM Handbook, Vol. 3: Alloy Phase Diagrams*, ASM International, 1992.

[16]  B. D. Cullity & C. D. Graham, *Introduction to Magnetic Materials,* 2nd Ed., IEEE Inc.

[17]  S.V. Vonsovskii, *Magnetism*, v. 2, Wiley (1974).

[18]  Sen Yang *et al.*, *Phys. Rev.* B 78 (2008) 174427.

[19]  R.M. Bozorth, *Ferromagnetism*, Van Nostrand (1951).

# The Physical Nature of "Giant" Magnetocaloric and Electrocaloric Effects

Physical nature of "giant" magnetocaloric and electrocaloric effects, MCE and ECE, is explained in terms of the new fundamentals of phase transitions, ferromagnetism and ferroelectricity. It is the *latent heat* of structural (nucleation-and-growth) phase transitions from a normal crystal state to the orientation-disordered crystal (ODC) state where the constituent particles are engaged in thermal rotation. The ferromagnetism or ferroelectricity of the material provides the capability to *trigger* the structural phase transition by application, accordingly, of magnetic or electric field.

American Journal of Materials Science 2013, 3(5): 105-109
DOI: 10.5923/j.materials.20130305.01

# The Physical Nature of "Giant" Magnetocaloric and Electrocaloric Effects

**Vitaly J. Vodyanoy[1,*], Yuri Mnyukh[2]**

[1]Biosensors Laboratory, Auburn University, Auburn Alabama USA
[2]Chemistry Department and Radiation and Solid State Laboratory, New York University, New York, NY USA

**Abstract**  Physical nature of "giant" magnetocaloric and electrocaloric effects, MCE and ECE, is explained in terms of the new fundamentals of phase transitions, ferromagnetism and ferroelectricity. It is the *latent heat* of structural (nucleation-and-growth) phase transitions from a normal crystal state to the orientation-disordered crystal (ODC) state where the constituent particles are engaged in thermal rotation. The ferromagnetism or ferroelectricity of the material provides the capability to *trigger* the structural phase transition by application, accordingly, of magnetic or electric field.

**Keywords**  Magnetocaloric, Electrocaloric, Ferromagnetic, Paramagnetic, Phase Transition, Latent Heat, Lambda Anomaly, Heat Capacity, Nucleation, Hysteresis, Molecular Rotation

## 1. Introduction

*Magnetocaloric effect* (MCE) is the heat emanated or absorbed when magnetic field is applied to magnetic material. In principle, some thermal effect can be produced by any magnetic material: the applied magnetic field makes its structure unstable, creating conditions for changing the currently existing directions of its spins toward the direction of magnetic field. If the structural rearrangement occurs, the energy gain turns to heat. The effect, however, is small. The term MCE usually designates the much stronger effect, observed when a *phase transition* in the material is involved.

Oliveira and von Ranke (O&R) have published a comprehensive review on MCE, with 238 references[1], which is a good representative of the theoretical literature on the subject. Their conclusion that "underlying physics behind the magnetocaloric effect is not yet completely understood" was an understatement. In fact, the whole search for physical origin of the phenomenon was misdirected. As a result of the conventional incorrect interpretation of ferromagnetic state and solid-state phase transitions in general, the MCEs were erroneously ascribed to changing in magnetization. We will show that, instead, it is rooted in changing of the *crystal structure*.

The purpose of the present article is to reveal the physical nature of the "giant" MCE and its electrical counterpart *Electrocaloric Effect* (ECE). Since the previous research was misdirected, so were the efforts to find the most effective

refrigerants for technologically sound magnetic refrigeration. In a seaming accordance with the term "magnetocaloric", the efforts were based on the belief that MCE is a *change in magnetic entropy* that can be estimated from magnetization measurements.

The MCE temperature is frequently reported as located in vicinity of magnetic phase transitions. The transitions were identified either as first order, or second order, or structural, or magnetic, or magnetostructural. There are descriptions of MCE as resulted from a "randomization of domains" at the Curie temperature. But the randomization process was not sufficiently understood either. Besides, the Curie temperature assumes a *second order* phase transition and, therefore, zero hysteresis, but hysteresis is known as a problem in the magnetic refrigeration technique. The MCEs in most experimental works were related to first-order phase transitions. The real molecular mechanism of the first order phase transitions could become a clue to the MCE origin, but they were treated only in a theoretically-formal manner, basically as "jumps" in the physical properties. The physical mechanism of these phase transitions remained in the dark.

The nature of MCE can be revealed only in terms of the real physics of a ferromagnetic state and solid-state phase transitions. The reader can find it in the book[2] and articles[3-8]. The next several sections are a background necessary for the explanation of the MCE and ECE that follows.

## 2. Classification of Phase Transitions by First- and Second-Order[2,3,9]

The classification of phase transitions by *first order* and *second order* has been taken for granted in the solid-state

* Corresponding author:
vodyavi@gmail.com (Vitaly J. Vodyanoy)
Published online at http://journal.sapub.org/materials

physics. It overwhelms the O&R review where it is used simply as statements "[X] is undergoing a first order phase transition", and "[Y] is undergoing a second order phase transition". In that sorting out, the "first order" were mentioned 76 times, and the "second order" 24 times. The only mentioned criterion was whether the entropy is a continuous or discontinuous function of temperature and magnetic field. If the former, it is a second-order transition, if the latter, it is a first order transition. It was not specified what *physical* process stays behind each of these two names. The theories suggested by O&R were not applicable to first order phase transitions. What makes phase transitions to be first or second order remained unknown.

The problem of the first/second order classification has been detrimental to solid-state physics for many decades. Presently it has simple solution: it should never be put forward. Soon after *second order* phase transitions were theoretically proposed in 1930th, M. von Laue and other prominent physicists rejected the possibility of their existence on thermodynamic reasons. In disregard of the objections, L. Landau introduced his theory of *second order* phase transitions, suggesting that they "may also exist". An analysis of the issue has led us to conclude that in reality *they do not exist*. The Landau's examples of second-order phase transitions turned out first order, as did the ferromagnetic transitions in Fe, Co and Ni, as do the transitions in all ferromagnetics and ferroelectrics, as do all "order-disorder" phase transitions. Not a single well-documented second-order phase transition exists. All current "second order phase transitions" are classified superficially and will ultimately be re-classified, so the classification itself will be *de facto* nullified. *All solid-state phase transitions materialize by a nucleation-and-growth rearrangement of the crystal structure.* This process is the most energy-efficient, requiring energy to relocate one molecule at a time, and not the myriads molecules at a time as a cooperative (second order) process requires.

## 3. Ferromagnetic State and Phase Transitions[5,10]

It has been experimentally established that MCE is tightly bound to ferromagnetic phase transition. All the current literature on the subject, including the O&R article, is based on the idea that MCE resulted from a "change in magnetic entropy". The ferromagnetic transition is deemed to be a change in the magnetic order in the same crystal lattice, even when the transition is identified as first order.

This interpretation of ferromagnetic phase transitions made discovery of the MCE origin impossible. Spin interaction in ferromagnetic material must be very strong in order to infer the "change in magnetic entropy" large enough to fit the largest MCEs actually observed. The Heisenberg's quantum-mechanical theory of ferromagnetism seemingly provided that strong interaction, called *electron exchange interaction*. The theory was to explain why a ferromagnet is

stable, while magnetic interaction of its spins, according to calculations, is not. The theory has been taken for granted, even though its initial verifications had to prevent its acceptance. The verifications produced a *wrong sign* of the exchange forces. In other words, the stability of ferromagnetic state would *decrease* rather than increase. Besides, the theory failed in many other respects. The sign problem was later carefully reexamined and found unavoidable. It was predicted that the "neglect of the sign may hide important physics"[11].

The predicted important physics resides in the power of crystal field. Its ability of imposing one or another magnetic order was overlooked. There is no need in the additional spin interaction to explain ferromagnetism. The real magnetic interaction constitutes only a few percentage points of the total free energy of a ferromagnetic crystal, the main part of it being the energy of crystal chemical bonding. The ferromagnetic crystal is stable *in spite* of the destabilizing effect of the magnetic spin interaction[5,12].

For years ferromagnetic phase transitions were regarded "structureless" by theoretical physicists. However, in time, experimental evidence was mounting that change in magnetic ordering in many cases is "accompanied by" a change in crystal structure. Such phase transitions are presently called "magnetostructural". The history of the matter needs in following two corrections: 1. Not only *many*, but *all* ferromagnetic phase transitions are "magnetostructural", and 2. The 'magnetic' and 'structural' roles in ferromagnetic phase transitions are actually different, considering that it is not the new magnetic ordering that brings about change of the structural change, as previously believed, but quite the reverse.

In relation to the search for the cause of MCE, it can be concluded that the thermal effect during a ferromagnetic phase transition comes from the *latent heat of the structural rearrangement*, while the contribution from the magnetic reordering is minor. Yet, that latter contribution might still be sufficient to determine which of the two competing crystal structures is preferable under given conditions.

## 4. Hysteresis, Nucleation and Range of Transition[6,13]

Hysteresis is always a hurdle in the way of practical realization of magnetic refrigeration. This fact is an indication that physical origins of MCE and hysteresis are closely related. Therefore, the statements by O&H "The present discussion does not consider the hysteresis effect" and "A microscopical description of the hysteresis effect on the magnetocaloric properties is still lacking" had to question their theory describing hysteresis-free second-order phase transitions. But hysteresis is always found. Its observation, even a smallest one, *binds the MCE to the process of nucleation and growth*.

At present, hysteresis has a detailed physical explanation. Not entering into all details, it is as follows. Magnetic ordering results from a *structural* rearrangement during ferromagnetic phase transition. The rearrangement materializes by *nucleation and growth*. The *nucleation* is not the classical fluctuation-based process described in textbooks. In a given crystal it is pre-determined. The nuclei are located in specific crystal defects - microcavities of a certain optimum size. These defects already contain individual information on the temperatures $T_n$ of their activation. The nucleation lags $\Delta T_n = T_n - T_0$ (at $T_0$ the free energies of the polymorphs are equal) are inevitable and not the same in different defects. These lags are the only cause of the *structural hysteresis*. Since each crystal structure determines its pattern of magnetic ordering, this structural hysteresis is a magnetic hysteresis as well. It is the hysteresis in phase transitions. The hysteresis of (re)magnetization by applied magnetic field has the same cause, namely, nucleation lags of the underlying structural rearrangements [10].

Considering that almost all real systems feature multiple nucleation, the phase transitions spread over a *range of transition* – temperature or magnetic depending on the driving parameter. The range as a whole is a subject to hysteresis, which means that the ranges in the direct and the reverse runs will not overlap.

# 5. Heat Capacity and λ-Anomaly[7,14]

MCE is victim of a mistake permeated all the literature where a heat capacity over the temperature region of phase transition is reported. Assigning the peaks like one in Fig. 1 to a heat capacity is erroneous. As an example, more than 30 such "specific heat" peaks were reproduced in[15]. These peaks have long history. After first recorded in 1922, they were named "heat capacity λ-anomalies" and regarded to be a feature of second order phase transitions. Their shape was meticulously analyzed in the efforts to figure out the nature of "critical phenomena". The "λ-anomalies" remained a mystery even after discovery that they "also" appear in first-order phase transitions. At that point, it seems, it had to become clear that the peaks are not an anomaly, but rather a *latent heat* of the structural phase transitions. That did not happen, possibly on the following reasons: (a) The first-order phase transitions were treated only formally, something like second-order phase transitions interrupted by "jumps", (b) The λ-peaks in the $C_P$ (T) plots in ferromagnetic phase transitions were ascribed to the magnetic properties and treated as a "specific heat near the Curie transition", and (c) The adiabatic calorimetry typically used in the measurements has a usually unnoticed limitation. An adiabatic calorimeter permits measurements only upon heating. While it correctly shows the λ-peak being endothermic, it is incapable to reveal that the same peak will be exothermic (looking downward) and located at lower temperature in the cooling run. This can be revealed with a differential scanning calorimeter and thus prove that the mysterious "anomaly" is simply a latent heat of the phase transition.

# 6. Crystal State with Rotating Molecules[16]

It is presently known that large MCEs are associated with the "order-disorder" phase transitions between magnetically ordered ferromagnetic and magnetically disordered paramagnetic states. As to the latter, the curiosity in the MCE research did not go farther of noting that spin orientations are randomized by thermal motion. While describing the MCE as change in magnetic entropy, it would be wise for the MCE research to recall about existence of orientation-disordered crystals with *thermal rotation of their molecules* and take a close look at that state.

This specific solid state was attracting a greater attention in the middle of the last century, culminating in the publication of the book[16], due to its novelty and hope to replenish the dwindling quantity of second-order phase transitions,. The substances revealing this state usually have rounded molecules. The state is characterized by the long-distance translation order in the molecular positions but not orientations owing to molecular thermal rotation. A rotational phase is always the highest on the temperature scale, right before melting. The rotation is never quite free, even when the long-distance orientational order is completely lost. With increasing temperature the molecular orientational disordering frequently occurs in several stages - from increased librations, to frequent jumps between certain discrete orientations, to 2-D rotation, to "free" rotation. However, even in the latter case the rotation is still hindered, retaining some degree of a short-distance orientational order.

All said about the whole molecules can be repeated when only their parts are subjected to thermal rotation. For example, in $NH_4Dy(SO_4) \cdot 4H_2O$ the following major stages were found[17]: (1) the ammonium ions are undergoing well defined librations at $95 < T < 200$ K, (2) their molecular motions at $200 < T < 275$ K are complex, probably a superposition of large-amplitude reorientational motion and small-amplitude librational fluctuations, and (3) they attain almost free rotation at $T > 275$ K.

All the order-disorder phase transitions in question, including those between the orientation-disordered stages, are first order, meaning they occur by nucleation and growth. They are among the solid-state phase transitions with the largest latent heat.

# 7. Paramagnetic State

Importance of correct interpretation of paramagnetic state goes far beyond of the MCE research. As for the literature on MCE, it is mentioned only as a state with orientation - disordered *spins* or, at best, as a state with the spins randomized by thermal motion. What is missing in that interpretation is that spins are not independent entities: they are bound to their atomic and molecular carriers (see Section 3). Therefore, a common statement like "thermal energy overcomes the interaction energy between the spins" is incorrect at the point that happens to be crucial for finding the MCE origin. The interaction energy between the spins is weak as compared to the molecular chemical bonding. It is the latter that the thermal energy must overcome to let the atomic-molecular spin carriers become rotating.

The truth of the matter is that paramagnetic crystals are the *orientation-disordered crystals (ODCs) with thermal rotation of their atoms and molecules*. The crystal restructuring in a *ferromagnetic − paramagnetic* phase transition should not be overlooked. As all *order − disorder* phase transitions, they occur by nucleation and growth and, therefore. involve latent heat.

## 8. Physical Nature of the "Giant" MCE in $Gd_5(Si_2Ge_2)$

The discovery of a "giant" MCE in $Gd_5(Si_2Ge_2)$ by Pecharsky and Gschneidner (P&G)[18] was a milestone toward development of the new refrigeration technology. But the discovery, best represented by the plot in Fig. 1, left the MCE without correct explanation of its physical nature. The MCE, represented by the λ-peak at the temperature of first-order phase transition 276 K, was ascribed to the " magnetic entropy change".

In terms of properties of the real solid-state phase transitions described in Sections 2 - 7, the nature of the "giant" MCE in question becomes almost self-explanatory. There is no basis for that MCE to be a "change of magnetic entropy". It is of *crystal-structural* origin, being the *latent heat of the structural phase transition* (by nucleation and growth) in the magnetic material. The MCE appears on the experimental heat capacity curves $C_P(T)$ as a peak that used to be a "λ-anomaly" indicative of the second-order phase transition. In the MCE studies these peaks were turned to a "heat capacity due to change in magnetic entropy". In fact, they are equally observed in non-magnetic materials and are proven[14] to be the *latent heat* of nucleation-and-growth phase transitions. In case of any doubts this can be verified by a differential scanning calorimetry: the λ-peak in the cooling run will emerge exothermic (looking downward) and, at that, at a lower temperature due to hysteresis.

It becomes now possible to resolve another contradiction in the interpretation by P&G[18] of their giant MCE (see Fig. 1). To be in accord with the empirically established rule, this giant MCE had to be ascribed to the small "anomaly" (arbitrarily claimed to mark a second order phase transition) produced by the *ferromagnetic − paramagnetic* (FM−PM) transition at 299 K − the type of phase transition where spin randomization or ordering occurs and the "magnetic entropy change" should be maximal. But the MCE was instead presented by the λ-peak produced by the (FM−FM) transition at 276 K, the transition type where both phases were found magnetically ordered.

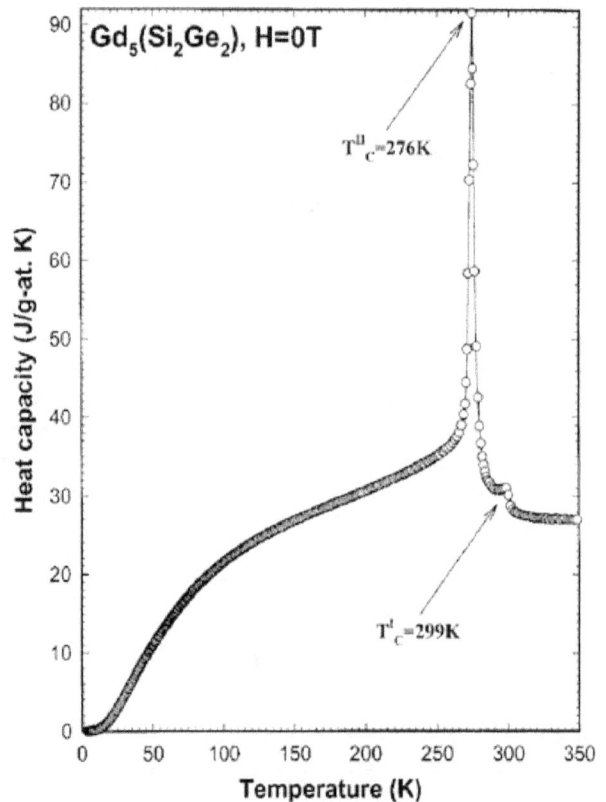

**Figure 1.** The zero magnetic field heat capacity of $Gd_5(Si_2Ge_2)$ from 3.5 to 350 K. The arrows point to heat capacity anomaly due to a second order paramagnetic ↔ ferromagnetic (I) transformation at 299 K and a first order ferromagnetic (I) ↔ferromagnetic (II) transition at 276 K. (This is the original figure capitation reproduced from[6]. Used with permission from APS Associate Publisher)

To explain this, we will trace the structural changes from the higher temperatures down. Molecular rotation in the ODCs is known to be hindered rather than quite free (see Section 6). The PM→FM phase transition at 299 K is $ODC_1 \rightarrow ODC_2$. The increased hindrance to molecular rotation with decreasing temperature turns the isotropic distribution of molecular orientations (and their spins) to anisotropic, thus converting the paramagnetic $ODC_1$ to a ferromagnetic $ODC_2$. Since it is still a *rotational − rotational* transition, only small change in the total crystal energy is involved (small latent heat peak at 299 K). The phase transition at 276 K features a giant MCE, for it is a transition of the rotational $ODC_2$ (FM) phase to the normal crystal with fixed molecular orientations. The latent heat of this phase transition is represented by the area of the λ-peak superimposed on the true $C_P(T)$ curve.

The magnetic change does not contribute essentially into the MCE. The magnetic component of the crystal free energy is small, being only sufficient to change the energy balance toward or against the ODC phase and *trigger* the phase transition when magnetic field is applied. The phase transition can also be triggered by change of temperature T, and the same latent heat could be named "TCE", or it can be triggered by pressure P and named "PCE". In fact, the λ-peak in Fig. 1 *is* the TCE. But using a temperature change as a trigger in the refrigeration technique is impractical. However, application of electric field to *ferroelectric - paraelectric* phase transitions is not impractical. It can produce the *electrocaloric effect* (ECE) quite analogous to the MCE.

# 9. Electrocaloric Effect (ECE)

As all solid-state phase transitions, *ferroelectric - paraelectric* phase transitions materialize by nucleation and growth. If triggered by application of electric field, their latent heat becomes *an electrocaloric effect ECE*. The ECE has long scientific history[19]. Accounting for its origin had turned out even more problematic, considering that nothing analogous to the "magnetic entropy change" and "electron exchange field" is applied to dielectrics. No sound explanation of the ECE exists. Yet, the ECEs comparable to the "giant" MCEs were attributed to the large *polarization change*. "near or above" the *ferroelectric → paraelectric* transition. Though random orientation of the electric dipoles in the paraelectric phase was recognized, the understanding was missing that this phase is an ODC where (dipolar) *molecules* are engaged in thermal rotation. It is the energy of the crystal restructuring, involving a conversion to thermal rotation of the constituent molecules, that makes accounting for the "giant" ECEs easy and identical to that for MCE.

# 10. Inverse MCE

Historically, the name "magnetocaloric effect" MCE was given to the *heating* upon application of magnetic field H. We now explain this as exothermic effect of the *disordered* (DIS) → *ordered* (ORD) structural phase transition caused by the orienting action the applied magnetic field H exerts on the disordered spins.

This left the name "Inverse MCE" to the endothermic effect also observed sometimes upon H application. Obviously, it results from the ORD → DIS transition.

In other words, application of H under some circumstances would destabilize the ORD phase more than it does to the DIS phase. In one such situation the action of the applied H can sufficiently strengthen spin interaction in the ferromagnet where that interaction, by itself is destabilizing (see Section 7). Another possible case is the H destroying antiferromagnetic order in the (AFM)ORD → DIS. transition.

# 11. Conclusions

■ Hopefully, the physical nature of the magnetocaloric and electrocaloric effects is not an enigma any more. It is always good to understand the origin of a physical phenomenon, but this is especially true in attempts to use it in practice. In case of MCE and ECE it may be now expedient to modify the direction of the search for best refrigerants. In the erroneous belief that magnetic properties are responsible for the *size* of MCE, a disproportional attention was paid to the magnetic measurements. In fact, ferromagnetism provides only the ability to *trigger* phase transitions by magnetic field. It may be prudent to shift a part of that attention toward analyzing the structure and properties of the orientation-disorder crystals (ODC) with thermal molecular rotation.

■ As for the ECE, it should be useful to know that search for its separate explanation can be canceled. The physical origin of the both MCE and ECE is the same, and it is *latent heat* of the structural phase transitions.

■ The function of the applied electric field is quite analogous to the applied magnetic field in the MCE case, namely, to *trigger* those phase transitions.

# REFERENCES

[1] N. A. de Oliveira, P. J. von Ranke, Phys. Reports 489 (2010) 89-159.

[2] Y. Mnyukh, Fundamentals of Solid-State Phase Transitions, Ferromagnetism and Ferroelectricity, Authorhouse, 2001[or 2nd (2010) Edition].

[3] Y. Mnyukh, Amer. J. Cond. Mat. Phys. 2013, 3(2): 25.

[4] Y. Mnyukh, Amer. J. Cond. Mat. Phys. 2013, 3(4):

[5] Y. Mnyukh, Amer. J. Cond. Mat. Phys. 2012, 2(5): 109.

[6] Y. Mnyukh, arxiv.org/abs/1103.2194.

[7] Y. Mnyukh, arxiv.org/abs/1104.4637

[8] Y. Mnyukh, arxiv.org/abs/1105.4299

[9] In Ref. 2, Sections 1.2 - 1.4.

[10] In Ref. 2, Chapter 4.

[11] J. H. Samson, Phys. Rev. B 51 (1995) 223-233.

[12] In Ref.2, Section 1.9.

[13] In Ref.2, Sections 2.5, 2.6, 4.13.

[14] In Ref. 2, Chapter 3.

[15] N.G. Parsonage, L. A. K. Staveley, Disorder in Crystals, Clarendon Press (1978).

[16] The plastically crystalline state, J. N. Sherwood (ed),John Wiley and Sons, (1979).

[17] S. Jasty, V. M. Malhotra, J. Phys. Cond. Mat. 4 (1992) 9469-9480.

[18] V. K. Pecharsky, K. A. Gschneidner, Phys. Rev. Lett. 78 (1997) 4494-4497.

[19] S. G. Lu, Q. M. Zhang, Advanced Materials 21 (2009) 1983-1987.

# The Nature of Ferroelectricity

The physical nature of ferroelectricity is intimately associated with the molecular mechanism of phase transitions. The conventional theory linking ferroelectricity with "soft-mode" phase transitions is invalid, since phase transitions do not materialize that way. Previous attempts to explain ferroelectricity could not be successful without detailed knowledge of the universal nucleation-and-growth mechanism of solid-state phase transitions, especially its epitaxial kind. Therefore, this mechanism is described again in the article. Its application to ferroelectricity has been done initially in the 2001 book Fundamentals of Solid-State Phase Transitions, Ferromagnetism and Ferroelectricity, but this time it is done in more complete form. The origin of ferroelectrics, their classification, formation of their domain structure, their stability, difference from pyroelectrics, formation of their hysteresis loops, parallelism with ferromagnetism, and more, are presented coherently in terms of the universal nucleation-and-growth mechanism of phase transitions.

American Journal of Condensed Matter Physics 2020, 10(1): 18-29
DOI: 10.5923/j.ajcmp.20201001.03

# The Nature of Ferroelectricity

Yuri Mnyukh

76 Peggy Lane, Farmington, CT USA

**Abstract**   The physical nature of ferroelectricity is intimately associated with the molecular mechanism of phase transitions. The conventional theory linking ferroelectricity with "soft-mode" phase transitions is invalid, since phase transitions do not materialize that way. Previous attempts to explain ferroelectricity could not be successful without detailed knowledge of the universal nucleation-and-growth mechanism of solid-state phase transitions, especially its *epitaxial* kind. Therefore, this mechanism is described again in the article. Its application to ferroelectricity has been done initially in the 2001 book *Fundamentals of Solid-State Phase Transitions, Ferromagnetism and Ferroelectricity*, but this time it is done in more complete form. The origin of ferroelectrics, their classification, formation of their domain structure, their stability, difference from pyroelectrics, formation of their hysteresis loops, parallelism with ferromagnetism, and more, are presented coherently in terms of the universal nucleation-and-growth mechanism of phase transitions.

**Keywords**   Ferroelectrics, Paraelectrics, Pyroelectrics, Spontaneous polarization, Phase transitions, Domain structure, Hysteresis loop, Soft-mode, Nucleation-and-growth mechanism

## 1. Introduction

The phenomenon of ferroelectricity is very similar to ferromagnetism. Ferroelectric crystals are spontaneously polarized crystals with the ability to change the direction of polarization by external electric field. A ferroelectric crystal phase usually (but not necessarily) arises by means of a phase transition from the non-polar higher-temperature *paraelectric* phase. There are two kinds of the paraelectric phases. Their zero polarization results either from a centrosymmetric distribution of positive and negative charges in the crystal unit cell, or from thermal molecular rotation in the crystal lattice. All ferroelectrics exhibit a *domain structure*. As the temperature goes down, ferroelectrics typically exhibit a few ferroelectric-to-ferroelectric phase transitions. These features distinguish ferroelectrics from *pyroelectrics* which have neither domain structure, nor succession of phase transitions, and their polarization cannot be changed by external electric field.

Since discovered in 1920 in Rochelle salt, ferroelectricity was identified in numerous other substances and presently has important technological applications. The literature is plentiful, but usually phenomenological or technological. The available physical theory of ferroelectricity, that

assumes a ferroelectric phase transition to be a *critical* phenomenon is invalid. We will show that a sound theory could not be developed without acceptance of the *universal nucleation-and-growth* molecular mechanism of solid-state phase transitions. This mechanism was put forward and described by the present author a number of times [*e. g.,* 1-3], but had not yet entered the orbit of conventional theoretical thinking. Considering that understanding of solid-state phase transitions is a key to physics of ferroelectricity, this universal mechanism is concisely described here again.

## 2. Universal Nucleation-and-Growth Mechanism of Phase Transitions

### 2.1. History

The correct interpretation of solid-state phase transitions is vital in revealing the physical nature of ferroelectricity, as it is in the cases of ferromagnetism [1,4] and superconductivity [5]. But there was a long-standing hurdle to overcome in the way of their investigation. In 1933 a theory of phase transitions took a wrong turn when Ehrenfest misinterpreted the high peak in the calorimetric measurements of liquid helium as evidence of a new type of phase transitions. He named them a *second-order* phase transitions, while the "usual" ones became *first order*. In spite of the rigorous proof by Justy and von Laue that the "second-order" phase transitions do not comply with thermodynamics, the suggested new phase transitions were welcomed by theoretical physicists as an opportunity to apply their knowledge of statistical mechanics.

In 1937 the research went further off course when L.

Landau presented the theory of *second-order* phase transitions. After being unanimously accepted for many years by physical science, such phase transitions were ultimately proven to not exist [5,6]. Until that fact is accepted by modern physics, they still have to be dealt with in the areas where phase transitions play important part. Ferroelectricity is one of these areas.

## 2.2. First- and Second-Order Phase Transitions

*Second-order* phase transitions are assumed to be a process when thermal fluctuations over the crystal are intensifying with the temperature until reaching a "critical point" $T_C$ where the crystal becomes totally unstable and the phase transition starts and completes at once. This dramatic event seemed to require changing the crystal structure instantly. However, such a macroscopic physical change at a "critical point" would be at variance with thermodynamics. Therefore, Landau made the transition to be "continuous": any changes of the crystal structure at $T_C$ are infinitesimal, only the crystal symmetry "jumps" sharply; the two phases cannot coexist; the temperature hysteresis is strictly zero; no heat is absorbed or emanated. These clear-cut characteristics have become instrumental in the eventual conclusion [1,5] that second-order phase transitions do not exist and made it easy to prove in any particular case that the ferroelectric phase transition is not *second order*.

*First-order* phase transitions are the ones we deal with in reality. Landau described them as exhibiting abrupt "jumps" in the state of the matter, at which latent heat is absorbed or released, symmetries of the phases are not related, and overheating or overcooling is possible. Other than listing their characteristics, no idea about the possible molecular mechanism of *first-order* phase transitions was offered.

Two crystal phases of a substance are independent of each other, whether their structures are similar or not. They occupy different areas in the phase diagram and can in principle be produced separately (from a liquid phase, for example) under the conditions where each is stable. Only one way of phase transition between them is in compliance with thermodynamics: an *infinitesimal* change of a controlling parameter (temperature, electric field, magnetic field, or pressure) gives rise to the material transfer of an *infinitesimal quantity* of the original phase to the new phase [5,7,8]. It is imperative to realize that no other way exists. Thus, a *cooperative* phase transition by instant "distortion" or "deformation" of the original crystal, or "displacement" of any its constituent particles, is not possible.

The *nucleation-and-growth* molecular mechanism described in next sub-sections is in compliance with the above requirements. It is *the only* alternative to the "second order" type. Its macroscopic indicators are: *interfaces, phase coexistence, range of transition, hysteresis*. Observation of any of them is sufficient to identify the transition as not "second order" and, therefore, statistical mechanics is not applicable, for there is no bulk statistics.

The real ("first order") and imagined "second order" phase transitions are antipodal and irreconcilable.

Any changes in solids are never a coordinated simultaneous movements of many particles. The real process is a molecule-by-molecule rearrangement; it is universal. The popular maneuver to justify treatment of some first-order transitions by statistical-mechanics by claiming they are "weakly first order" (or "almost," or "close to" second order) is nonsensical: either the phase transition is localized at interfaces, or involves simultaneously all particles in the bulk. No middle is imaginable.

## 2.3. Nucleation in Microcavities. Hysteresis

Nucleation of the alternative phase in solid-state phase transitions differs in all respects from the theoretically-born fluctuation-based statistical picture described in the Landau and Lifshitz textbook (Section 150 in [9]). Nucleation in a given crystal is pre-determined as for its location and temperature. It would not occur at all in perfect crystals, but such crystals are extremely rare. The nucleation sites are located in specific crystal defects – microcavities and microcracks of a certain optimum size. These sites contain information on the condition (e. g., temperature) of their activation. Nucleation temperatures are not the same in different defects. A nucleus is activated only *after* the temperature $T_0$ marking the equality of free energies $F1 = F2$ has been passed. The nucleation lags are inevitable in both directions of phase transitions. *Hysteresis is a necessary component of the phase transition mechanism.* At least some finite hysteresis is inescapable. Its value is not exactly reproducible. The "ideal phase transition" (i. e., without hysteresis) cannot occur due to absence of a thermodynamic driving force $\Delta F$. ($F1 - F2 = 0$ at $T = T_0$).

Along with the hysteresis of initiation of the new phase there is also much smaller, but consequential, hysteresis of sustaining the transition when the two phases already coexist. It will be explained in section 2.5.

## 2.4. Temperature of Phase Transition. Range of Transition

Temperature of phase transitions is regularly treated in scientific literature as $T_C$, calling it "critical temperature" (or "Curie temperature" or "critical point") – the term designated for "critical" phenomena. This is a consequential mistake, a source of countless misinterpretations of the nature of phase transitions. Then, when the value of a measured property $p$ shows a rapid change "in the vicinity of transition" on the experimental plots $p(T)$, this is interpreted as a pre-transition effect. Or the plots may show $p(T)$ not being sharp, and in some cases even rather smooth. The phase transitions look complicated. Location of the phase transition temperature becomes a subject of guessing, and the curves become a subject of mathematical delineation.

The cause of these seeming complications is that a phase

transition cannot occur at $T_0$ due to hysteresis. It is always spread over some *temperature range* located *after* the $T_0$ has been passed. In that range the coexisting phases continue to be static over their bulk, while their ratio is changing by local rearrangement at the interfaces. Simply marking the correct $T_0$ position (it can be found by extrapolation) will erase the mystery. As for the transitions looking sharp in the plots $p(T)$, they will not be quite sharp if recorded with a higher resolution.

## 2.5. Rearrangement at *Contact* Interface

A solid-state phase transition is intrinsically a *local* process. It proceeds by "molecule-by-molecule" structural rearrangement at interfaces only, while the bulk of both the original and emerged phases remain static (Fig.1). No "jumps" of macroscopic quantities or properties occur. The seeming "jumps" appearing on the experimental recordings are simply the differences between the crystal structures or their physical properties when the transition range is either narrow, or passed quickly, or both. Coexistence of the phases during phase transition is self-evident. Simple observation of simultaneously present phases is a solid proof of the nucleation-and-growth phase transition.

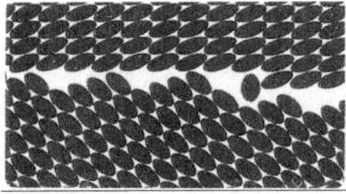

**Figure 1.** Stepwise (or edgewise) molecule-by-molecule rearrangement at the *contact* interface. The process is located at "kinks" on the crystal faces of the emerging phase (upper part) and proceeds by building up layer-by-layer that way, while the original phase is dissolving. No structural orientation relationship between the phases is necessary. This molecular mechanism is in accord with thermodynamics requiring only an "infinitesimal" quantity (in macroscopic terms) of the matter to be transferred at a time

The phase transition illustrated by Fig. 1 leads to the question of whether it is *continuous* or *discontinuous*. Real transitions are not discontinuous in the traditional interpretation of first-order phase transitions; they are not an instant jump of the whole crystal into a new phase. Instead, the transition is quantitatively spread over a temperature interval. But within that interval it is locally discontinuous. Every completed molecular layer requires nucleation to initiate the next layer. We call it 2-D nucleation as compared to the 3-D nucleation that initiates a new phase. The 2-D nucleation lags give rise to hysteresis, and they are also responsible for the Barkhausen effect (Section 4.10 in [1]). The hysteresis of this type is much smaller than the one described in section 2.3 but is essential, for example, in the formation of ferroelectric domain structures.

## 2.6. Epitaxial Type of Phase Transitions

Fig. 1 illustrates a case when the crystal orientations of the phases are not related. This is because the nuclei

initiating the phase transition in the crystal defects can, in principle, be oriented arbitrarily. However, a structural *orientation relationship* (*OR*) is frequently observed. That does not mean these transitions materialize by a modification or deformation of the original structure, as is still generally believed. These transitions materialize by nucleation and growth, no matter how similar the crystal structures could be. The two phases are independent of each other. The cause of the *OR* in these cases is *oriented nucleation*.

When *OR* is rigorous, we call the transition *epitaxial*. There are two circumstances when it takes place. One is in pronounced layered structures. A layered crystal consists of strongly bound, energetically advantageous two-dimensional units − molecular layers, usually appearing almost unchanged in both phases. When comparing the structures, they differ only by their layer stacking. The interlayer interaction in these crystals is relatively weak by definition. Accordingly, the divergence between the total free energies of the two structural variants is small, which is a precursor of polymorphism. Nucleation occurs in the tiny interlayer cracks. Given the close structural similarity of the layers in the two polymorphs, the nucleation is *epitaxial* due to the orienting effect of the "substrate" − the opposite surface of the crack (Fig. 2). The layers after the transition retain their original orientation.

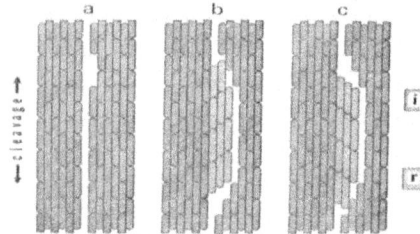

**Figure 2.** Initial stage of phase transition in a layered crystal. Orientation of the layers is retained. (a) A microcrack: its flat shape enhances the surface interaction and is responsible for the orienting effect. (b,c) Alternatively oriented embryos of the new phase consisting of the same layers, but different layer stacking. After coming into contact, a domain structure is produced

Another case of *epitaxial* nucleation is when the unit cell parameters of the polymorphs are extremely close (roughly less than 1%) even in non-layered crystals. That condition is met, for example, in some *order-disorder* phase transitions (as observed in *paramagnetic − ferromagnetic* phase transition of iron [1,10]).

*Epitaxial* phase transitions exhibit themselves in a specific way: the OR is rigorous, the Laue X-ray patterns of the phases appear almost identical, temperature hysteresis can be small and easily missed, so is the latent heat, *etc*. It is the *epitaxial* phase transitions that are taken for "displacive", "soft-mode", or "second-order". In reality they are all "first order," having materialized by nucleation and growth. All ferroelectric phase transitions, sorted out by "displacive" and "order-disorder" in the literature, have one and the same molecular mechanism: *epitaxial nucleation-and-growth*.

# 3. "Classical Principles" of Ferroelectrics

## 3.1. Identifying Those Principles

The book *Principles and Applications of Ferroelectrics and Related Materials* by Lines and Glass (L&G), 677 pages, including 30 pages of references, first published in 1977, was re-printed in 2001 by Oxford University Press in its series *Oxford Classical Texts in the Physical Sciences* [11]. We will examine whether that stamp of approval by a major publisher was justified. First, it should be found what particular principles were behind the words "principles of ferroelectrics". Then we consider if they are valid.

In a few decades after the discovery of ferroelectricity the attempts to understand the phenomenon were aimed at analyzing "ferroelectric instability" in every particular case that ferroelectricity was found. It was believed that spontaneously polarized crystals appear by *displacive* phase transitions of the "original" (or "prototype") non-polar *paraelectric* crystal phases. The positive and negative charges in paraelectrics neutralize each other, and ferroelectric transitions represent *displacements* of some their atoms or molecules. However, the actual physical process of the displacements remained unknown. We find that the above "principles" by L&G are reduced to their claim that those displacements have a "*soft-mode*" molecular mechanism: "The microscopic breakthrough came in 1960 with the recognition of the fundamental relationship between lattice dynamics and ferroelectricity and, most importantly, of the existence of soft-mode instability at a ferroelectric transition … The great value of the soft-mode concept is that it enables us to form a reasonably unified microscopic picture of ferroelectricity … It then becomes possible to... grasp the relationship of ferroelectricity to the more general field of structural transitions and even to critical phenomena in general". Accordingly, the book was saturated by "critical phenomena": soft-modes, second-order phase transitions, Hamiltonians, Curie points, statistical mechanics, Landau theory, and so on. Nothing of that has the slightest relation to the real physics of ferroelectricity.

Remarkably, all those "critical phenomena" coexisted with the recognition that "most ferroelectric phase transitions are not of second order but first." Lack of knowledge in the area of phase transitions (characteristic of the most current literature as well) made the L&G contribution to the physical sciences counterproductive.

## 3.2. The Soft-Mode Concept

Over the years there was a sequence of proposed mechanisms of phase transitions: displacive, martensitic, topological, soft-mode, incommensurate, scaling, quantum. None was successful [8]. The *soft-mode* concept was put forward in about 1960 to explain the mechanism of *displacive* ferroelectric transitions, but then artificially extended to the non-displacive *order-disorder* ferroelectric transitions and, finally, to all "structural" phase transitions.

According to the theory, a structural phase transition is a cooperative *distortion* of the original crystal structure as a result of atomic / molecular shifts (displacements) in it. This distortion is produced by one of the "soft" (*i. e.*, low-frequency) optical modes. That soft mode further "softens" with decreasing temperature until reaching $T_C$ – the critical point when its wavelength becomes comparable with the crystal parameters. At that point the atomic displacements make the crystal unstable and the displacements suddenly become "frozen", producing the alternative phase. It is to be noted that such transition, considered to be second order, is not "continuous", as second-order phase transitions are defined to be.

The soft-mode concept was developed, tested and demonstrated by using ferroelectric $BaTiO_3$ as an example; even "jumps" of the physical properties at the (non-existent) Curie points were calculated. But the same $BaTiO_3$ was used by Landau to illustrate a *continuous* second-order phase transition. Evidently, at least one of the two conflicting approaches must be incorrect. Actually, both were wrong: the "first-order" nature of the $BaTiO_3$ phase transitions is now well established. Instead of the fixed critical points, temperature hysteresis of the transition temperatures wider than 10°C was observed.

In many experimental works on phase transitions no "good" soft-modes, properly softening toward the transition temperature, have been found. Many phase transitions were between the crystal structures so different that presenting the process as "distortion" of the original structure was unimaginable. Also, the model was inconsistent with the observations of phase transitions by moving interfaces. But the best argument against the soft-mode mechanism is the photographs of well-shaped crystals growing inside the original single crystal [1,12]. The soft-mode concept of phase transitions has not justified the hopes of its inventors. Except for some ferroelectric theories, it has fallen into oblivion.

## 3.3. Thirty Years After "Soft-Mode Principle"

A collection of eight articles entitled *Physics of Ferroelectrics – A Modern Perspective* [13] was intended to update the physical science on ferroelectricity 30 years after the 1977 L&G book. The Editors, Rabe, Ahn and Triscone, called the L&G book "classic" and a "sound foundation." The articles were supposed to be "a complement that brings the reader from that sound foundation up to the present". As for the physics of ferroelectricity, the following excerpts from the introductory article by K. M. Rabe and four co-authors, entitled *Modern Physics of Ferroelectrics: Essential Background,* tell all we need to know about the promised scientific advancement:

*"The symmetry-breaking relation between the high-symmetry paraelectric structure and the ferroelectric structure is consistent with a second-order transition, and can be described with a Landau theory. Phonon spectroscopy continues to play a central role in the characterization of ferroelectric transitions. The soft-mode*

*theory is illuminating, despite the fact that in many perovskite ferroelectrics the transition is weakly first order".*

Thus, there was no positive development in understanding of ferroelectricity. Application of the Landau theory was postulated without any justification. The "soft-mode" concept was not discarded. The science was still on the wrong *critical phenomenon* track. The *nucleation-and-growth* approach to phase transitions was still absent. The only new element was labeling many first-order phase transitions "weakly" as an excuse to justify treating them as second order. Not surprisingly, all basic facts about ferroelectrics remained explained incorrectly, or not explained at all.

### 3.4. I.S. Zheludev's Comprehensive Review

The earliest comprehensive review of the phenomenon of ferroelectricity belongs to I. S. Zheludev, who presented it in two 1971 volumes *Physics of Crystalline Dielectrics* [14] and the 1973 Russian version *"Foundations of Segnetoelectricity"* [15]. That pioneering work, preceding the L&G "classical" book by several years, was hardly possible to overlook; yet it was not included in their 30-page list of references. The Zheludev work, however, was more reasonable and informative as far as physics of ferroelectricity is concerned. All aspects of ferroelectricity were systematically investigated and discussed: crystallography, definition, difference from pyroelectrics, crystal structure of ferroelectrics, their stability, their domain structure, domain boundaries, dependence of spontaneous polarization on temperature, hysteresis loops of polarization, and more.

It is not to say Zheludev was always correct. For the most part he was not. His efforts were impaired by a disarray in the adjacent scientific area – phase transitions. There was a unanimous belief in the existence of second-order phase transitions, as well as in the *displacive* mechanism, and Zheludev did not escape it. He sorted out the ferroelectric phase transitions by first/second order and treated the formation of ferroelectrics from paraelectric phase as "displacements" and "ordering." The *universal nucleation-and-growth mechanism* was only in its formulation stage at that time and certainly was unknown to Zheludev. His ultimate goal to uncover the physical origin of ferroelectricity was not achievable. But his work represented a good starting point for moving forward. He will be mentioned time and again in next sections.

# 4. Foundations of Ferroelectricity (Our Presentation)

### 4.1. On Classification of Ferroelectrics

Kittel [16] stated that ferroelectric crystals are classified into *displacive* and *order-disorder* "according to whether the transition is associated with the ordering of ions or .. the

displacements of a whole sublattice of ions of one type...". Zheludev explained that it was impossible to provide a satisfactory classification of ferroelectrics in terms of their properties, but "[that] can be made on the basis of the mechanism of the appearance of spontaneous polarization", namely, was it by "displacements of certain ions" or by "ordering of certain structure elements".

There was a silent problem with that classification: it was incomplete. Some ferroelectrics did not have the "Curie point," because, as was explained by several authors, the material melts before the Curie point was reached. It was hardly a sound argument, for it tells us only about a *disappearance* of the ferroelectrics with rising temperature. Reverse the direction of temperature change – and we deal with formation of these ferroelectrics by melt crystallization. Therefore, one more type had to be added to the *displacive* and *order-disorder* ferroelectrics: *melt-crystallized* ferroelectrics. After that, to make the classification complete, the ferroelectrics crystallized from solutions should be added to the three.

The current classification of ferroelectrics is based on the lingering ignorance about solid-state phase transitions. Formation of ferroelectric crystals by two different mechanisms of phase transitions was the only base for that classification. But such mechanisms do not exist – and neither does the base. There is only one mechanism for all ferroelectrics to appear: *crystallization.*

### 4.2. Formation of Ferroelectrics

Lines and Glass stated that "the great value of the soft-mode concept is that it enables us to form a reasonably unified microscopic picture of ferroelectricity..." Instead, it rather divided ferroelectrics in two categories, since the soft-mode concept covered only the *displacive,* but not the *order-disorder* type. And it does not even count the ferroelectrics that appear from liquid phase.

Ferroelectrics do not emerge by displacements in paraelectric crystals or by ordering. Phase transitions involving a *modification* of the original crystal do not exist. The essence of the *universal nucleation-and-growth mechanism,* described in section 2, is a construction of the new crystal from the molecular material supplied by the original crystal. It is not like a renovation of an old house; it is a construction of the new one nearby by using bricks from the old house. The new house can look different; but it can also be very similar, in which case it would not mean that it appeared by a "distortion" of the old one.

The structure of the original phase in ferroelectric phase transitions, and whether it is solid or liquid, is irrelevant. The soft-mode concept did not provide the unified picture of how ferroelectricity comes into being. Crystallization does. All ferroelectrics (and antiferroelectrics too) are produced by nucleation and growth, using the original phase as a source of building material whatever it is: ordered crystal, orientation-disordered crystal, or liquid.

### 4.3. Distortion and Strain

According to the conventional theories, a *displacive* ferroelectric phase transition is *distortion* of the nonpolar paraelectric crystal by displacement of certain particles in its unit cell. The distortion causes *strains*, so they were added to the theory. Both these phenomena turn out to be fictitious after the realization that real phase transition – crystallization – does not involve systemic distortion of the ferroelectric crystals it produces. It is to be noted that the ferroelectric crystals grown from liquids cannot be thought of as distorted and strained original crystals, because the original medium was liquid phase.

### 4.4. Stability of Spontaneously Polarized Crystals

Spontaneously polarized crystals appear in the 10 polar symmetry classes. The question is why they appear at all, considering that a system of the mutually neutralized positive and negative electric charges, as in paraelectrc crystals, is more thermodynamical advantageous than a system of parallel dipoles in polar crystals. Therefore, Zheludev pointed out: "It is essential to understand, at least qualitatively, why the spontaneous polarization states of some crystals are thermodynamical stable" [15]. The answer, which typically escaped attention, is rooted in the fact that the interaction of electric charges alone in polar crystals does not determine their overall stability. The total contribution of all components of the crystal free energy $F$ does.

Potential energy of atomic or intermolecular bonding always towers above other $F$ components. This is the case even in molecular crystals where, as known, the Van der Waals' intermolecular bonding is relatively weak. As estimated by Kitaigorodskii [17], the energy of electrostatic interactions in organic crystals does not exceed 5% of the sublimation energy. Thus, the contribution of the dipole-dipole interaction to the free energy $F$ in spontaneously polarized molecular crystals is small. The relative contribution must be even much smaller in the crystals with a chemical bonding. The part played by such a "small contributor" is not always negligible, but is not essential in the polar crystals. *They are stable in spite of the disadvantage of a parallel alignment of their electric dipoles.* Quite similar reason made the Heisenberg's theory of ferromagnetism unnecessary [1,4], which was an attempt to find the cause of stability of a ferromagnetic state .

### 4.5. Origin and Formation of Ferroelectric Domain Structures

#### 4.5.1. The Unifying Factor

The underlying factor unifying the origin of all ferroelectrics is a very close geometrical proximity of unit cells of their polymorphs. In case of crystallization from a paraelectric phase, both ordered and disordered, this proximity makes the phase transitions *epitaxial*. The domains appear in certain crystallographic orientations relative to the original phase, and therefore, to each other – a necessary condition of a domain structure formation (Fig.2).

In the melt-crystallized ferroelectrics the "substrate" for oriented nucleation does not exist initially. However, in certain cases the first crystals that appear from the liquid phase, especially when they have a layered structure, can play its role.

#### 4.5.2. Primary Cause: Multinucleation in Epitaxial Transitions

The origin of ferroelectric domain structure is a key to understanding ferroelectricity. One of the Zheludev's explanations was that "splitting into domains occurs … because the polarization, which arises at different points in a crystal at the Curie temperature is equally likely to assume any of the various crystallographically equivalent directions". Basically, that was right idea, but in its formulation the *effect*, which is polarization, should be replaced by the *cause*, which is nucleation.

**Table 1.**   Unit cell of the polymorphs of 1,2,4,5-tetrachlorobenzene (TCB) and aniline hydrobromide (AHB)

| | | T C B | | AHB | |
|---|---|---|---|---|---|
| | | $C_6H_2Cl_4$ | | $C_6H_5NH_3Br$ | |
| $T_o$, °C | | -60 | | +22 | |
| Phase | | Low-temperature | High-temperature | Low-temperature | High-temperature |
| | | | | | |
| Symmetry | | Triclinic | Monoclinic | Monoclinic | Rhombic |
| Unit cell (Å) | $a$ | 9.60 | 9.73 | 16.725 | 16.77 |
| | $b$ | 10.59 | 10.63 | 5.95 | 6.05 |
| | $c$ | 3.76 | 3.86 | 6.81 | 6.86 |
| | □ | 95° | (90°) | | |
| | □ | 102.5° | 103.5° | 91°22' | (90°) |
| | γ | 92.5° | (90°) | | |

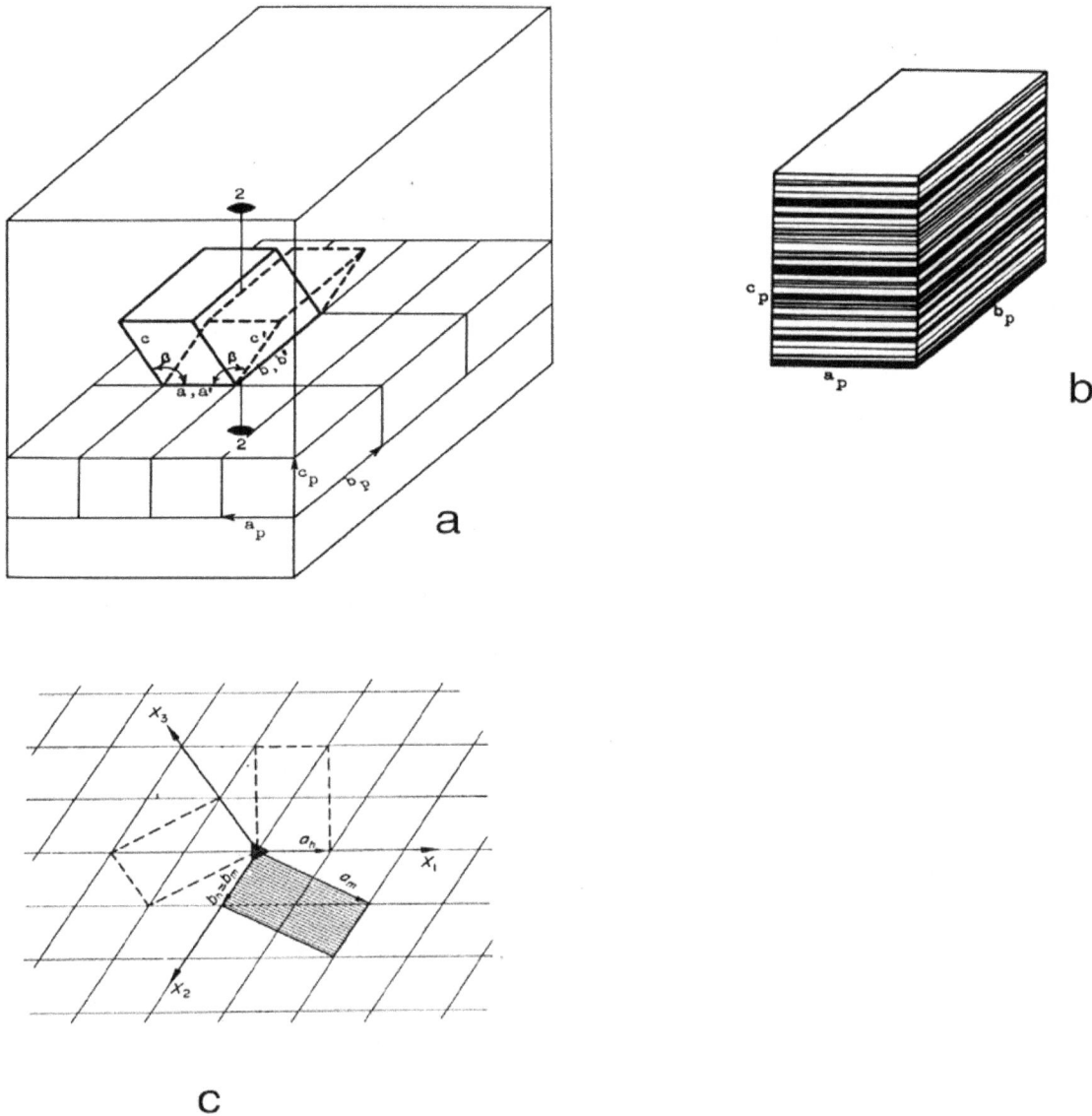

(a) Two equally probable orientations of a monoclinic lattice of the resultant phase in the original crystal of a rectangular lattice and cleavage (*001*).
(b) Growth of each nucleus leads to formation of a lamina of one particular orientation of the two possible. The arising new phase is becoming a laminar structure of the domains of two alternating fixed orientations shown as black and white. (Two neighboring laminae of the same orientation merge into one.)
(c) A three-orientation case represented by *s*-triazine $(CH)_3N_3$. Transition from the high-temperature rhombohedral lattice to the monoclinic gives rise to three, rather than two, equally probable domain orientations (one is shaded, two others are shown by dashed lines). The subscripts *m* and *h* are related to the monoclinic and hexagonal unit sells respectively. The final laminar structure consists of the domains of the three orientations alternating at random.

**Figure 3.**   To the formation of "twin" domain structure in epitaxial phase transitions as a result of oriented multi-nucleation of a lower-symmetry phase. The nuclei assume two or more equally probable orientations

Formation of the domain structures becomes less mysterious after we realize that it is not unique to ferroelectrics. We can temporarily set aside the possible effect of polarization and look into the cases of some non-polar substances that have domain structure. Two substances presented in Table 1 are not ferroelectrics. In both cases the lower symmetry low-temperature phase is reported to appear after phase transition in two opposite orientations in approximately equal quantities. That phenomenon was interpreted as the consequence of a *displacive* mechanism acting in different directions in different parts of the crystal. But such mechanism does not exist. The universal nucleation-and-growth mechanism has to be invoked.

The very close geometrical proximity of the crystal unit cells of the polymorphs, as seen in Table 1, assures that the phase transitions are *epitaxial*. Analysis of the molecular packing indicated that their crystals have layered structure. Formation of the layered domain structures is almost inevitable in epitaxial transitions if the emerging phase has a lower symmetry. In practice, phase transitions in layered crystals are always multinuclear. Even a seemingly perfect single crystal of a layered structure contains multiple submicroscopic wedge-like chinks parallel to the cleavage planes. These tiny chinks act as the nucleation sites from which the layered domains of the new phase grow. As illustrated in Fig. 3, an oriented nucleus can appear with equal probability under two orientations related to one another as crystallographic twins.

The two crystals can be brought into coincidence with a two-fold axis perpendicular to (*001*). Growth of a nucleus gives rise to the formation of a *laminar domain* of one or the other orientation. When the sides of the adjacent domains of the same orientation meet, they merge into a single domain. Thickness of the domains varies. A layered domain structure sketched in Fig. 3b emerges with strict alternation of the orientations. Any two adjacent domains are crystallographic twins.

This mechanism is the primary cause of the domain structure in ferroelectrics; it is not related to their polarization.

### 4.5.3. Zigzagging by Little Mistakes

After a nucleus of ferroelectric phase appears in the layered paraelectric "matrix," its growth gives rise to a new layered structure with almost unchanged geometrical parameters of the layers. At every stage of that growth the new layers can serve as substrates for the 2-D nucleation of subsequent layers in one or the other alternative orientation. This time the two orientations are not equally probable: the one which would continue the domain growth is more advantageous than the one starting a new domain. But their nucleation energies are so close that "mistaken" nucleation acts do occur from time to time. The result of such "zigzagging" (Fig.4) is a structure of thinner domains.

### 4.5.4. "Head-to-Tail" Packing

The described process of "zigzagging" is intensified by the ferroelectric polarity. It makes the "mistaken" nucleation more frequent by providing electrostatic dipole attraction ("head-to-tail", see Fig. 4) on the sides of a domain interface.

While the growth of the domain is still preferable, the difference between the two nucleation energies is further equalized. The very fine division of ferroelectric crystals into domains, found in experimental studies, becomes more understandable.

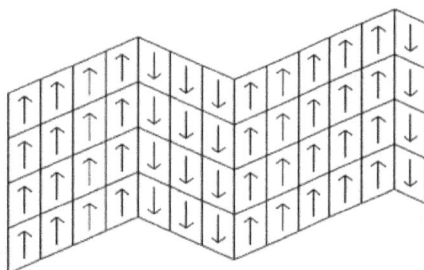

**Figure 4.**  Ziggzaging effect in formation of ferroelectric domain structure due to the 2-D nucleation errors

### 4.5.5. The "Wall" of Zero Thickness

The term "domain wall" was adopted by ferroelectricity from ferromagnetism. Conventional science claimed that ferromagnetic domains are separated by a thick "wall" of hundreds of lattice periods in which spins gradually turn to the new direction. The conclusion was inferred from the Heisenberg ferromagnetic theory. That theory is now proven invalid [1,4]; the ferromagnetic domain interfaces are geometrical twin boundaries. Zheludev believed in those thick ferromagnetic domain boundaries, but ferromagnetism was not the area of his expertise. Nevertheless, he arrived at the different, almost correct, conclusion that ferroelectric domain walls "should – in most cases – be of the order of the lattice constant".

The domain boundaries – in *all* cases – are *twin interfaces, i. e.* having zero thickness. Formation of a ferroelectric crystal and its domain structure is one and the same process. The two components of that formation, as described in the previous sections, already contain the answer to what a ferroelectric domain interface is. In the *multinucleation crystallization* (section 4.5.2), when the sides of the adjacent domains of the same orientation meet, they merge into a single domain. If they have different orientations, they form a geometrical twin interface of zero thickness. In the *zigzagging crystallization* (section 4.5.3), the perfect geometrical matching during 2-D epitaxial nucleation gives rise to a perfect twin interface as well.

As for the shape and directions of the domain interfaces, they are rational crystallographic flat faces of the domain single crystals when they grow.

## 4.6. What Differentiates Ferroelectrics from Pyroelectrics

### 4.6.1. Ferroelectrics

The direction of spontaneous polarization of ferroelectrics can be changed by an external electric field $E$ because they have domain structure. They have it, as explained in section 4.5.1, due to a very close geometrical proximity of their crystal unit cells to those of the paraelectric phase. Accordingly, their free energies are close too, $F_{par} \approx F_{fer}$. The domain structure consists of multiple single crystals of identical crystal structure arranged in alternating orientation and separated by twin interfaces. The free energy $F$ of the crystal lattice in all domains is the same until an electric field $E$ is applied. The field $E$ tilts the energy balance in favor of the domains with dipole moments $P$ closer to the $E$ direction. Every two neighboring domains now embody two competing crystal phases of slightly different $F$. The phases are separated by a very low potential barrier, considering that 3-D nucleation is not needed and the activation energy to move a twin interface is the lowest imaginable. A phase transition is initiated and proceeds by the molecule-by-molecule relocation at all the twin interfaces. The process is complete when the "unfavorable" domains disappear and all polarization vectors acquire the direction of the "preferred" domains. The system is now spontaneously polarized. Reversing the $E$ direction will reverse the process of molecular rearrangement, giving rise to what is called *polarization switching*. Evidently, this "switching" is not instant and can sometimes be a rather slow process.

## 4.6.2. Pyroelectrics

The direction of spontaneous polarization in ferroelectrics can be changed because they have a domain structure. Pyroelectrics do not have the domain structure and their polarization cannot be changed. It remains to be clarified why they do not have it and why polarization cannot be changed without it.

The reason why pyroelectrics do not have their domain structure is the same reason that most crystals in nature do not have it. It is because the specific conditions for its formation, discussed in sections 4.5.1 and 4.5.2, are not satisfied. The reason why applied electric field cannot change polarization without domain structure is best illustrated with molecular crystals, where closeness of molecular packing correlates with the crystal free energy, namely, the closer packing, the lower the free energy. How close the packing can be depends on shape of the molecules. Some molecules have two or more potential packing versions of almost equal density $\rho1 \approx \rho2 \approx \rho3...$, so their free energies $F1 \approx F2 \approx F3...$ . Their crystals are prone to exhibit polymorphism, and some may qualify to be ferroelectrics if they are polar and the second condition – closeness of the crystal unit sell parameters – is also satisfied. Other molecules have such a set of their closest packing that $\rho1 \ll \rho2, \rho3 ...$ , and, accordingly, $F1 \ll F2, F3...$ . So, the system with $F1$ rests on the bottom of a deep potential well, so the weak driving force of applied electric field cannot extricate it from there. Pyroelectrics belong to the latter category. Also, even if $F1 \approx F2$ in some cases, the system would remain pyroelectric when the crystal parameters of the phases are not very close.

# 5. Ferroelectric Hysteresis and Loops

## 5.1. Zheludev's Description

The dielectric hysteresis loops in an alternating electric field are the most prominent feature of ferroelectrics. Zheludev gave a description of such a loop (Fig. 5). Here it is in brief (translated from [15] with special attention to preserve the used terms).

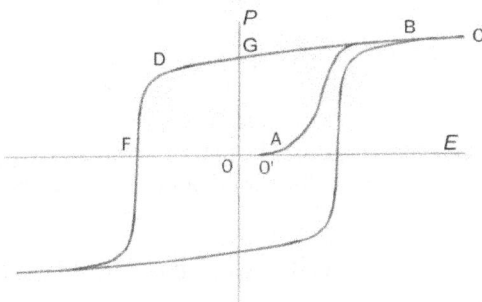

**Figure 5.**  Ferroelectric hysteresis loop $P$ $(E)$. $P$ – polarization, $E$-applied electric field

In the region O $\rightarrow$ A, the polarization $P$ is roughly proportional to field $E$; it is accompanied by an elastic (reversible) displacement of domain borders, or by no domain processes at all. Over region A $\rightarrow$ B, field $E$ makes $P$ grow at the expense of domain processes (emergence of domains and their growth). At point B the polarization, connected to the domains, attains saturation, and the crystal becomes a monodomain. The region B $\rightarrow$ C represents the linear relation between $P$ and $E$ and characterizes the induction polarization not caused by domain processes.

In lowering $E$ from point C, the polarization, starting from point B, does not follow the path of the initial polarization. It does not return to zero at $E = 0$, but retains nearly all its saturation value (point D). This means that its monodomain state is preserved almost entirely in the absence of an applied field. However, when $E$ approaches its coercivity (OF), the polarization drops precipitously, becomes zero and, reversing its polarity, reaches large values. "This means that at $E$ values close to coercivity an intensive repolarization of the domains is going on: changing their polarity sign to the opposite."

## 5.2. Discussion and Updating

It is essential to note that the description was written exclusively in terms of the *polarization* changes; the *structural* rearrangements at the interfaces were not discerned. The cause of polarization hysteresis remained unknown. The whole domain structure was misunderstood as a *single crystal* with different areas (domains) of one or another direction of polarization. The term "domain processes" was not specified. Together with the emergence of new domains and their growth, it also presumed a polarization change by domains as a whole. *The main shortcoming of the whole approach was that the underlying structural rearrangements were not perceived.*

The truth is that the domain structure is not a single crystal, but a conglomeration of *differently spatially oriented* single crystals. Therefore, rearrangement of the crystal structure between differently oriented crystals is the only way to reorganize the polarization. Structural rearrangement is the *cause*, and change in polarization is its *effect*.

It should be noted that a good understanding of the hysteresis and its loops could not be achieved without taking into account the specifics of nucleation and rearrangements at crystal interfaces in solid state. One relevant fact remained unknown in the literature as regarding the ferroelectric (and ferromagnetic) hysteresis loops: the same type of hysteresis loops has been observed in solid-state phase transitions when a physical property was measured *vs.* temperature T (Fig. 6). Their origin and formation is the key to explaining ferroelectric hysteresis loops. Their major characteristics are:

- They are reflecting the mass ratio of the two phases *vs.* T, regardless of the physical property measured;
- Nucleation lags are responsible for their formation and shape;
- Since all phase transitions in solids proceed by nucleation and growth, the hysteresis loop is an inalienable feature of any solid-state phase transition.

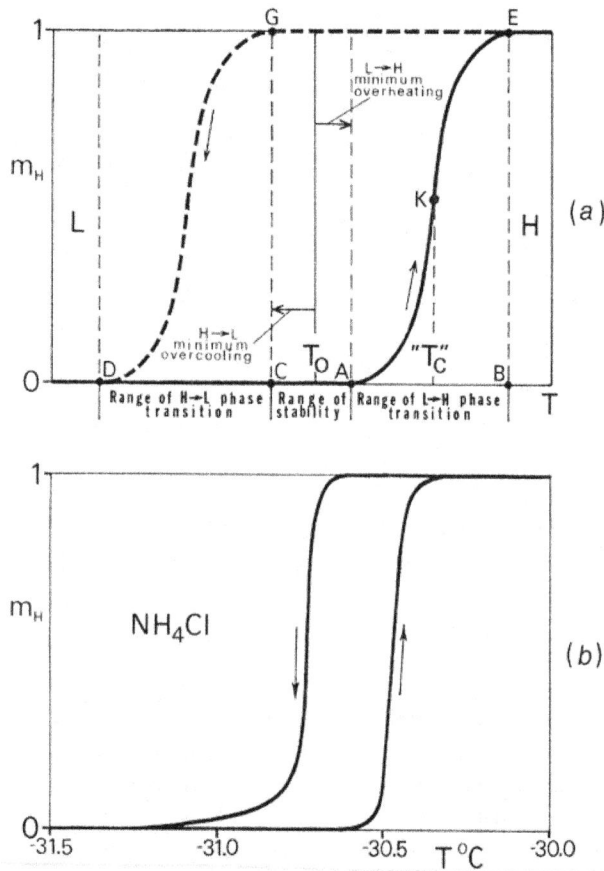

**Figure 6.** (a) (Schematic) Hysteresis loop $m_H(T)$ of a solid-state phase transition. Letters H and L denote high- and low-temperature phases. $m_H$ - mass fraction of H phase. The S-shaped curves AE and GD represent $m_H$ in the heterophase (L+H) temperature ranges of transition. Together they form a hysteresis loop DAEGD. Range of stability CA consists of two threshold lags too small to activate nucleation within. The inflection point K is not a "critical" (or Curie) point. It marks the temperature of the maximum number of activated nucleation sites; (b) (Experimental) Hysteresis loop of $NH_4Cl$ (Dinichert [15])

The similarity between ferroelectric and non-ferroelectric hysteresis loops is not accidental: they all represent a process of structural phase transition by nucleation and growth under the action of a variable. In the polymorphic transitions the variable was temperature, while in ferroelectric phase transitions it is the applied electric field. The changing temperature transfers the matter across the line in a phase diagram that separates its areas of stability and instability. It is the unstable state of the matter that is the driving force of a nucleation-and-growth phase transition. Similarly, application of electric field to the (differently spatially oriented) crystalline domains destabilizes those of the unfavorable polarity, thus causing nucleation and growth at the domain interfaces. It is quite appropriate to interpret that process as an E-driven structural phase transition where *quantities* of the participating phases (domain mass of each orientation) are measured in the units of polarization P.

### 5.3. Structural Interpretation of the Loop

Now we turn to Fig. 5 again to interpret the same loop in terms of nucleation-and-growth phase transitions. First, the original crystalline object at $E = 0$ calls for description. It is a system of alternatively oriented spontaneously polarized single-crystal domains (they will be distinguished as *right (r)* and *left (l)* phases) separated by twin interfaces – as it emerged after temperature phase transition from paraelectric phase. Their relative mass quantities are equal; $m_{r} = m_{l}$. The applied field E initiates phase transition $m_l \rightarrow m_r$, which proceeds by a growth of the r-phase at the expense of the l-phase at the existing interfaces. No "polarization switching" of the whole domains occurs. The transition starts from point O' rather than O due to hysteresis. Initially (O'→A) it is slow, but accelerates when the number of interfaces engaged in the process increases. After point A most interfaces are active and the $dm_r/dt$ quickly reaches its maximum at the inflection point of the O'–A–B curve. Then, when the process nears point B, it decelerates until it becoms zero at that point. The resultant curve O'→A→B acquires the typical S-shape characteristic of the processes when two factors work in the opposite directions. In this case they are the accelerating speed of structural rearrangement and the dwindling amount of remaining matter to be rearranged.

At point B the transition is completed, all the domains becoming r-oriented. The sample is now like a single crystal, but it contains numerous potential nucleation centers, both inherited and newly formed. Increasing E from point B (line B→C) induces some reversible intra-structural changes caused by elastic deformation of the dipoles themselves, That segment does not belong to the hysteresis loop, but reminds us that the B – G part of the loop contains a small reversible component. If the field E is disconnected at that point, the polar crystal of the P = (OG) value will remain stable for the reason explained in section 4.4. Decreasing E from point B directs P along B→G→D. In sufficiently good crystals it is almost a straight line, only slightly deviated from being horizontal due to the above-mentioned elastic reversible effect. The $m_r \rightarrow m_l$ phase transition will not start at point G: the field E of negative sign must first overcome the 3-D nucleation hysteresis of that transition. A mass nucleation and growth of the l-domains after point D gives rise to the high-speed phase transition $m_r \rightarrow m_l$ by the crystal rearrangements at all interfaces. (No polarization switching of the whole domains).

If the value of saturation polarization (point G) set to be $P_{max} = 1$, the plot in Fig. 5 can serve as a quantitative measure of the structural phase transitions between the *right*-oriented and *left*-oriented domains. The upward ordinate axis can now be marked $m_r$, and the downward ordinate marked $m_l$.

## 6. Parallelism with Ferromagnetism

In the article *Analogies and Differences between Ferroelectrics and Ferromagnets* [19] its author N. A. Spaldin maintained that "systematic comparison between the behavior of ferromagnets and ferroelectrics, does not, to our knowledge, exist in the literature." But, using Google and the key words 'ferromagnetism ferroelectricity', it would take just minutes to find the book [1] with these words in its title where that has been done; even the hysteresis loops of polarization and magnetization were presented as a single combined "ferroic" loop.

Spaldin noted that there are many similarities between ferroelectrics and ferromagnets, but concluded that they are "superficial." That conclusion was based on the alleged differences in the fundamental physics of spontaneous magnetization and spontaneous polarization, namely, that the former is established by the quantum-mechanical electron exchange field, while the latter resulted from a distortion of the non-polar prototype crystal phase.

The fact is, the theory of ferromagnetism in question is invalid, as is the assumption of a spontaneous polarization by crystal distortion. The ways the spontaneous ferromagnetic and ferroelectric states emerge are not different:

- Both ferromagnetic and ferroelectric crystals emerge by *nucleation-and-growth* phase transition from their para-phase.
- Both cases are those where the parameters of crystal unit cells of para- and ferro- phases are very close, which is a prerequisite for formation of domain structure.
- Spontaneously magnetized crystals emerge from the paramagnetic phase by *epitaxial* nucleation-and-growth phase transition [4], as do the spontaneously polarized crystals from the paraelectric phase.
- During those phase transitions a domain structure is formed. Its primary cause in both cases is multiple nucleation at random in different points of the para-matrix.
- The domain boundaries are twin interfaces in both cases.
- Both the magnetization and the polarization proceed exclusively by a molecule-by-molecule rearrangement at interfaces.
- The existence (*i. e.*, thermodynamic stability) of both spontaneously polarized and magnetized states is provided by crystal field.
- The hysteresis loops of polarization in electric field, $P$ $(E)$, and magnetization in magnetic field, $M$ $(H)$, are caused by the 3-D and 2-D nucleation lags in the process of structural rearrangement between the alternatively oriented domains.
- Both hysteresis loops, $P$ $(E)$ and $M$ $(H)$, represent quantitative rates of the structural phase transitions between the domains of different spatial orientation, even though the loops are recorded in $P$ and $M$ units.

## 7. Conclusions

The fundamental physics of ferromagnetism and ferroelectricity is identical. The difference is reduced only to the saturation values of spontaneous magnetization $M_S$ and spontaneous polarization $P_S$. Unlike the elementary magnetic dipole moments, which are a property of atomic spins, the electric charges of a ferroelectric dipole are spatially separated. Therefore, while $M_S$ by itself does not appreciably depend on $H$, $P_S$ depends on $E$ to some extent due to its sensitivity to the distance between the plus and minus charges. The additional polarization induced by the field $E$ is superimposed on that which is caused by the domain structural rearrangements and it is noticeably present on the hysteresis loops. Specifically, the saturation polarization $P_S$ $(E)$ continues to grow even after $P$ of all domains is parallel. This induced polarization is strictly reversible and has nothing to do with hysteresis. Beyond that effect, the crystal structure treats the electric and magnetic dipoles equally.

## REFERENCES

[1]   Y Mnyukh, Fundamentals of Solid-State Phase Transitions, Ferromagnetism and Ferroelectricity, Authorhouse (2001) [or $2^{nd}$ (2010) Edition].

[2]   Y Mnyukh, Molecular mechanism of polymorphic transitions, Molecular Crystals and Liquid Crystals 52, 163-200 (1979).

[3]   Y Mnyukh, Mechanism and kinetics of phase transitions and other reactions in solids, American Journal of Condensed Matter Physics 2013, 3(4): 89-103.

[4]   Y Mnyukh, Ferromagnetic state and phase transitions. Amerian Journal of Condensed Matter Physics 2(5), 109-115 (2012).

[5]   Y Mnyukh and V J Vodyanoy, Superconducting state and phase transitions, American Journal of Condensed Matter Physics 2017, 7(1): 17-32.

[6]   Y Mnyukh, Second-order phase transitions, L Landau and his successors, American Journal of Condensed Matter Physics 2013, 3(2): 25-30.

[7]   M Azbel, Preface to: R Brout, Phase Transitions (Russian ed.), Mir, Moscow (1967).

[8]   Y Mnyukh, On phase transitions that cannot materialize, American Journal of Condensed Matter Physics 2014, 4(1):1-12.

[9]   L D Landau and E M Lifshitz, Statistical Physics, Addison-Wesley (1969).

[10]  Y Mnyukh, The true cause of magnetostriction, American Journal of Condensed Matter Physics 2014, 4(3): 57-62.

[11]  M E Lines and A M Glass. Principles and Applications of Ferroelectrics and Related Materials, Clarendon Press (1977); Oxford University Press (2001).

[12]  Y Mnyukh and N N Petropavlov, Polymorphic transitions in molecular crystals – 1. Orientations of lattices and interfaces. Journal of Physics and Chemistry of Solids 33 (1972) 2079-2087.

[13]  Physics of Ferroelectrics – a Modern Perspective, K M Rabe, C. H Ahn, J-M Triscone (Eds.), Springer-Verlag, Berlin

(2007).

[14] I S Zheludev, Physics of Crystalline Dielectrics, v. 1 and 2 (1971).

[15] I S Zheludev, Foundations of Segnetoelectricity, Atomizdat, Moscow (Rus., 1973).

[16] C Kittel, Introduction to Solid State Physics, 4th Ed., Wiley, (1971).

[17] A I Kitaigiorodskii, Molecular Crystals and Molecules (Rus.), Academic Press (1973).

[18] P Dinichert, La transformation du NH4Cl observee par la diffraction des rayons X, Helvetica Physica Acta 15, 462-475 (1942).

[19] N A Spaldin, Analogies and Differences Between Ferroelectrics and Ferromagnets (pp. 175-218 in Ref. 13).

# Superconducting State and Phase Transitions

From the days when superconductivity was discovered its science was entangled by the unresolved problem of the relationship between superconducting state, its crystal structure and its phase transitions. The problem was exacerbated by the division of phase transition into first and second order. Adding to that the erroneous idea of the structural identity of the superconducting and normal phases, the whole issue was uncertain and confusing. This article shows that the phase transitions attested previously as a second order are first order by (a) demonstrating that the "heat capacity λ-anomaly" in liquid *He* was actually latent heat, (b) confirming the Justy and Laue thermodynamic consideration that the second-order type cannot exist, and (c) revealing that not a single well-proven second-order phase transition in solid state was ever found. Thermodynamic arguments are presented that being a first-order phase transition means crystal rearrangement, and that the universal *nucleation-and-growth* mechanism of phase transitions is the way it to materialize. In particular case of normal – superconducting phase transitions we have shown that (1) the normal and superconducting crystal structures are not identical, (2) the phase transition between them proceeds by a structural rearrangement, (3) the *epitaxial* version of the *nucleation-and-growth* is its molecular mechanism, (4) because the mechanism is universal toward all kinds of solid-state phase transitions, there was no basis to assume that it leads to description of superconductivity. Elimination of the identified wrong concepts and misinterpretations will accelerate the advancement of the science of superconductivity toward creation of efficient inexpensive high-temperature superconductors.

American Journal of Condensed Matter Physics 2017, 7(1): 17-32
DOI: 10.5923/j.ajcmp.20170701.03

# Superconducting State and Phase Transitions

**Yury Mnyukh[1,\*], Vitaly J. Vodyanoy[2]**

[1]Chemistry Department and Radiation and Solid State Laboratory, New York University, New York, NY, USA
[2]Biosensors Laboratory, Alabama Micro/Nano Science and Technology Center, Auburn University, Auburn Alabama, USA

**Abstract**    From the days when superconductivity was discovered its science was entangled by the unresolved problem of the relationship between superconducting state, its crystal structure and its phase transitions. The problem was exacerbated by the division of phase transition into first and second order. Adding to that the irroneous idea of the structural identity of the superconducting and normal phases, the whole issue was uncertain and confusing. This article shows that the phase transitions attested previously as a second order are first order by (a) demonstrating that the "heat capacity λ-anomaly" in liquid *He* was actually latent heat, (b) confirming the Justy and Laue thermodynamic consideration that the second-order type cannot exist, and (c) revealing that not a single well-proven second-order phase transition in solid state was ever found. Thermodynamic arguments are presented that being a first-order phase transition means crystal rearrangement, and that the universal *nucleation-and-growth* mechanism of phase transitions is the way it to materialize. In particular case of normal – superconducting phase transitions we have shown that (1) the normal and superconducting crystal structures are not identical, (2) the phase transition between them proceeds by a structural rearrangement, (3) the *epitaxial* version of the *nucleation-and-growth* is its molecular mechanism, (4) because the mechanism is universal toward all kinds of solid-state phase transitions, there was no basis to assume that it leads to description of superconductivity. Elimination of the identified wrong concepts and misinterpretations will accelerate the advancement of the science of superconductivity toward creation of efficient inexpensive high-temperature superconductors.

**Keywords**    Superconductivity, Crystal structure, Phase transition, First order, Second order, Weakly first order, Lambda-anomaly, Heat capacity, Latent heat, Ehrenfest, Laue, Structure distortion, Nucleation-and-growth

## 1. Introduction

Investigations of the crystal structure and phase transitions in the context of superconductivity basically follow two different lines. More frequently it is an analysis of the structural type, structural features and polymorphic forms of the known or potential superconductors, which they exhibit under different temperatures, pressures, substitutions or dopings. This line is aimed at empirical search for better superconductors. The second line is investigations of the normal – superconducting phase transitions at the temperature where they occur. The present article is of the latter type.

Immediately after the discovery of superconductivity in 1911 by Kamerlingh-Onnes the question on *how* superconducting state emerges from the higher-temperature "normal" non-superconducting state arose. Finding the answer was perceived to be a major step toward understanding the superconducting state itself. With that in mind, it had to be first established whether superconducting phase has its individual crystal structure. A science historian

P. F. Dahl described the relevant state of affairs of those days in his 1992 book [1]: "Onnes then returned to the crucial question raised first by Langevin at Brussels in 1911 regarding evidence for a change of phase at the transition point. An experimental search for a latent heat of transformation was planned by Onnes with the assistance of the American Leo Danna … in 1922-1923. … In any case, Onnes seems by now to have been more inclined to view the transition in terms of Bridgman's polymorphic change. This possibility was, however, negated by a decisive experiment concluded by Keesom shortly before the 1924 Solvay Conference. Analysis of Debye-Scherrer X-ray diffraction patterns revealed no change in the crystal lattice of lead …".

It will be shown in the present article that the conclusion about crystallographic identity of pre-superconducting and superconducting phases was in error. It had and is still has a lasting detrimental effect on the investigation of superconductivity. Its general acceptance relied on an experiment performed on only one superconductor without later verification with improving techniques and without validation on other superconductors. It served as a basis to erroneously categorize these transitions as "second order" after Ehrenfest [2] introduced his first/second order classification in 1933.

All superconductors known at that time are now recognized by most to exhibit first order phase transition, but

\* Corresponding author:
ymnyukh@hotmail.com (Yury Mnyukh)
Published online at http://journal.sapub.org/ajcmp

the newer ones are sorted out in both ways, and even hovering somewhere in between. Assigning the "order" is being done in a formal manner, using unreliable criteria, and without specifying the physical content behind each "order". Missing is a familiarity with the solid-state phase transitions in general, an adjacent area of solid- state physics on its own rights. Our purpose is to present the physical picture of a *normal – superconducting* phase transition and its relation to the superconducting state in terms of the general *nucleation-and-growth* molecular mechanism of structural rearrangements. This mechanism is described in Section 5 in sufficient detail. The "order" issue needs to be analyzed first.

## 2. First-Order and Second-Order Phase Transitions

What are a "phase" and a "phase transition"? *"A phase, in the solid state, is characterized by its structure. A solid-state phase transition is therefore a transition involving a change of structure, which can be specified in geometrical terms"* [3]. There are only two conceivable ways the phase transition may occur without being at variance with thermodynamics. An infinitesimal change of a controlling parameter (dT in case of temperature) may produce either (A) emerging of an infinitesimal quantity of the new phase with its structure and properties changed by finite values, or (B) a cooperative infinitesimal "qualitative" physical change throughout the whole macroscopic bulk [4]. It is imperative to realize that not any third way can exist. Thus, finite changes by "distortion" or "deformation" of the crystal, or "displacement" of its particles are not possible.

There is no guarantee that both versions 'A' and 'B' materialize in nature. However, in 1933 Ehrenfest formally classified phase transitions by *first-order* and *second-order* (see Section 3). The validity of the classification was disputed by Justi and Laue by stating that there is no thermodynamic or experimental justification for second-order phase transitions. Their objections were ignored and forgotten over the ensuing decades (see Section 6).

In 1935-1937 Landau [5, 6] developed the theory of *second-order* phase transitions. But he acknowledged that transitions between different crystal modifications are "usually" *first-order*, when "jump-like rearrangement takes place and state of the matter changes abruptly"; at that, a latent heat is absorbed or released, symmetries of the phases are not related and overheating or overcooling is "possible". As for *second-order* phase transitions, they "may also exist", but no incontrovertible evidence of their existence was presented. (It should be noted that the expression "may exist" implicitly allows something not to exist either). *Second-order* phase transitions must be cooperative and occur without overheating or overcooling at fixed "critical points" where only the crystal symmetry changes, but structural change is infinitesimal. It is specifically

emphasized that "second order phase transitions, as distinct from first-order transitions, are not accompanied by release or absorption of heat". These features were in accord with those by Ehrenfest, namely, no latent heat, no entropy change, no volume change, and no phase coexistence.

Since then it has become universally accepted that there are "jump-like discontinuous" *first-order* phase transitions, as well as "continuous" *second-order* phase transitions without "jumps". The latter fit well the thermodynamic requirement *'B'*, leaving *first-order* phase transitions to be associated with the requirement *'A'*.

The theoretical physicists were so preoccupied with second-order phase transitions, even trying to treat first-order transitions as second order that neglected to take a close look at the first-order transitions. If they did, the need to comply with the requirement *'A'* would be noticed. That requirement contains essential information on their molecular mechanism. In a sense, they are also "continuous". Contrary to the everybody's conviction, instant jump-like macroscopic changes do not occur. The transitions proceed by successive transfer of *infinitesimal* amounts of the material from initial to the resulting phase, which on the microscopic scale means molecule-by-molecule. During that process the two phases inevitably coexist. It will be shown in Section 4 that the experimentally discovered mechanism exactly fits the requirement *'A'*.

Without going into the Landau theory itself, there were several shortcomings in its presentation:

(a) He had not answered the arguments of the contemporaries, M. Laue among them, that second-order phase transitions do not - and cannot – exist.

(b) The only examples used in illustration of second-order phase transitions, $NH_4Cl$ and $BaTiO_3$, turned out to be *first order*.

(c) The theory was unable to explain the so called "heat capacity λ-anomalies" which were regarded being second order. The fact that they also appeared in first-order phase transitions was not addressed either.

(d) Overheating and overcooling in first-order phase transitions are not only "possible", but inevitable, considering that they will not proceed when the free energies $F_1$ and $F_2$ are equal and, therefore, the driving force is absent.

(e) First-order phase transitions were described as "abrupt jumps-like rearrangements", apparently assuming them to comprise a finite bulk of matter at a time. The same is seen from allowing first-order phase transitions to sometimes occur without hysteresis, in which cases they would occur at the single temperature point $T_0$ when $F_1 = F_2$, involving instantly the whole crystal – in contradiction with the condition *'A'*.

(f) When claiming that second-order phase transitions "may also exist", no estimates, or even arguments, were presented that they could be more energy advantageous in some cases. However, a strong argument can be raised that they can never materialize, considering that the first-order phase transitions (matching the condition 'A') would be always preferable, requiring energy to relocate only one molecule at a time, rather than the myriads of molecules at a time as a cooperative process of second-order phase transitions assumes.

# 3. Ehrenfest Classification: Construction on Quick Sand

### 3.1. "λ-Anomaly" in Liquid Helium and the Idea of Second-Order Phase Transitions

In 1932 Keesom and coworkers [7] discovered the "heat capacity λ-anomaly" in liquid He phase transition (Fig. 1). This apparent new type of phase transition was the reason for Ehrenfest to put forward his classification. Here are a few excerpts from the 1998 article [8] where the events were summarized:

"It is important to note a general scheme was generated on the basis of just one "unusual" case, liquid helium. It seemed probable that the scheme would be applicable to other known systems, such as superconductors … The liquid-helium lambda transition became one of the most important cases in the study of critical phenomena – its true (logarithmic) nature was not understood for more than ten years … The Ehrenfest scheme was then extended to include such singularities, most notably by A. Brain Pippard in 1957, with widespread acceptance. During the 1960's these logarithmic infinities were the focus of the investigation of "scaling" by Leo Kadanoff, B. Widom and others. By the 1970s, a radically simplified binary classification of phase transitions into "first-order" and "continuous" transitions was increasingly adopted." To the voluminous theoretical efforts listed in the last excerpt, the 1982 Nobel Prize to Kenneth Wilson for his scaling theory of second-order phase transitions had be added. Unfortunately, all that activity descends from a wrong interpretation of the "λ-anomaly" in liquid He: it is not heat capacity, but the latent heat of the first-order phase transition. Allegorically, the tower was erected on quicksand.

### 3.2. "Heat Capacity λ-Anomalies" in Solid-State Phase Transitions

It is not clear why Ehrenfest did not invoke the "heat capacity λ-anomaly" in $NH_4Cl$ phase transition, discovered by F. Simon in 1922 [9], for it exhibited the same peculiar λ-shape as that in liquid He. Later on, that λ-peak was reproduced many times by other workers and numerous similar cases were also reported. Thus, more than 30 experimental λ-peaks presented as "specific heat $C_P$ of

[substance] vs. temperature T" were shown in the 1978 book by Parsonage and Staveley [10]. Theorists were unable to account for the phenomenon. P.W. Anderson wrote [11]: "Landau, just before his death, nominated [lambda-anomalies] as the most important as yet unsolved problem in theoretical physics, and many of us agreed with him… Experimental observations of singular behavior at critical points… multiplied as years went on… For instance, it have been observed that magnetization of ferromagnets and antiferromagnets appeared to vanish roughly as $(T_C-T)^{1/3}$ near the Curie point, and that the λ-point had a roughly logarithmitic specific heat $(T-T_C)^0$ nominally". Feynman stated [12] that one of the challenges of theoretical physics today is to find an exact theoretical description of the character of the specific heat near the Curie transition - an intriguing problem which has not yet been solved.

The problem was hidden in the erroneous interpretation of available experimental data. There were three main reasons for that theoretical impasse. (1) Attention was not paid to the fact that λ-peaks were actually observed in first-, and not second-order phase transitions. (2) The first-order phase transitions exhibited latent heat, but it was mistaken for heat capacity. (3) An important limitation of the adiabatic calorimetry utilized in the measurements was unnoticed.

### 3.3. The "λ-Anomaly" in $NH_4Cl$ is Latent Heat of First-Order Phase Transition

The canonical case of "specific heat λ- anomaly" in $NH_4Cl$ around -30.6°C has been reexamined [13] (also Section 3.4.6 and Appendix 2 in [14]). That case is of a special significance. It was the first where a λ-peak in specific heat measurements through a solid-state phase transition was reported and the only example used by Landau in his original articles on the theory of continuous second-order phase transitions. This particular phase transition was a subject of numerous studies by different experimental techniques and considered most thoroughly investigated. In every calorimetric work (e. g. [15-21] a sharp λ-peak was recorded; neither author expressed doubts in its heat capacity nature. The transition has been designated as a cooperative order-disorder phase transition of the lambda type and used to exemplify such a type of phase transitions. No one claimed that the λ-anomaly was understood.

It should be noted that many of the mentioned calorimetric studies were undertaken well after 1942 when the experimental work by Dinichert [22] was published. His work revealed that the transition in $NH_4Cl$ was spread over a temperature range where only mass fractions $m_L$ and $m_H$ of the two distinct L (low-temperature) and H (high-temperature) coexisting phases were changing, producing "sigmoid"-shaped curves. The direct and reverse runs formed a hysteresis loop shown in Fig. 2a. The fact that the phase transition is first-order was incontrovertible, but not identified as such.

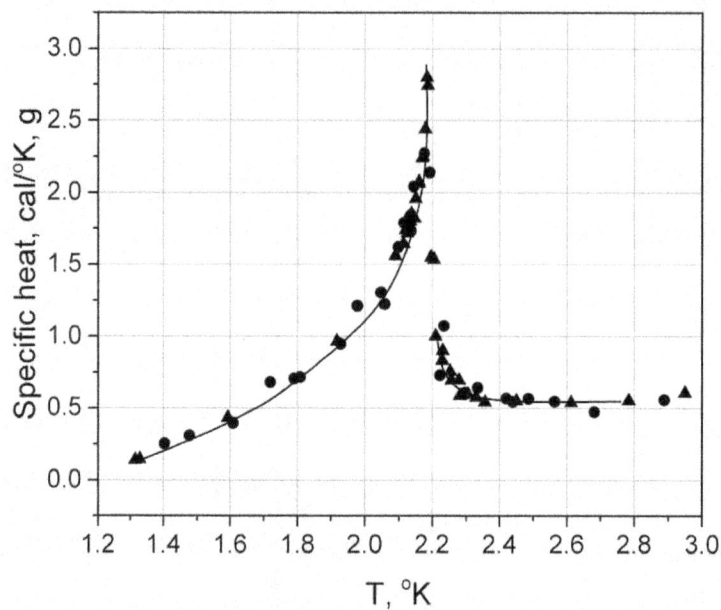

**Figure 1.**   The "λ-Anomaly" recorded in heat capacity measurements of liquid He phase transition (adopted from [7])

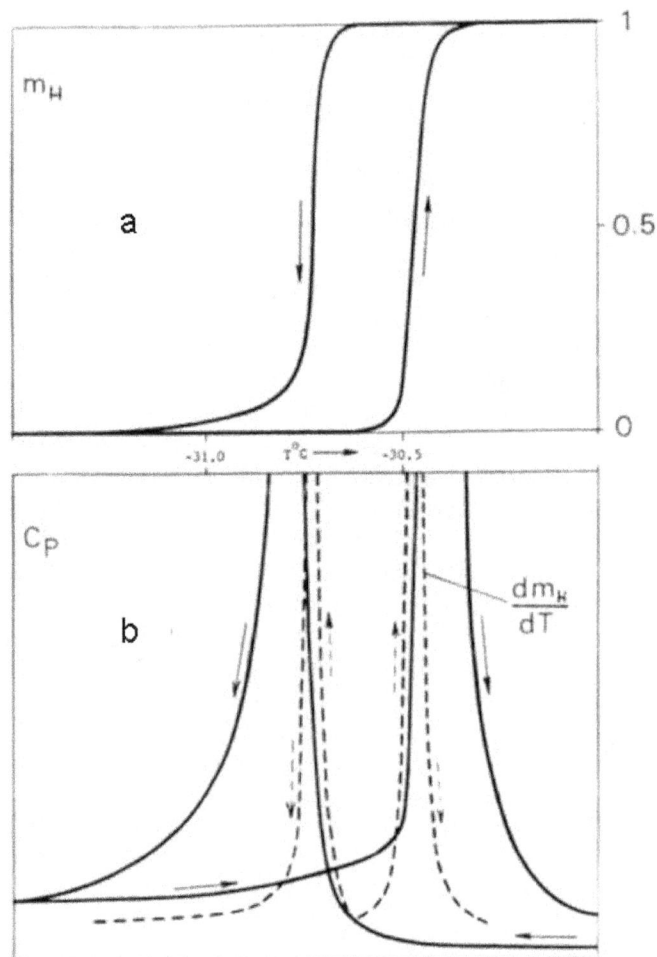

**Figure 2.**   Phase transition in NH₄Cl. a. The hysteresis loop by Dinichert [22] represents a mass fraction of high-temperature phase, $m_H$, in the two-phase, L+H, range of transition; $m_L+m_H =1$. b. Solid lines: The λ-peaks from calorimetric measurements by Extermann and Weigle [21]. The peaks are a subject of hysteresis. The plots 'a' and 'b' are positioned under one another on the same temperature scale to make it evident that the shape of the peaks is proportional to a first derivative of the $m_H(T)$ (dotted)

In Fig. 2b the Dinichert's data were compared with the calorimetric measurements by Extermann and Weigle [17]. The latter exhibited "anomalies of heat capacity" (as the authors called the λ-peaks) and the hysteresis of the λ-peaks. Because of the hysteresis, it had to be evident (but was not) that the λ-peaks cannot be of a heat capacity, considering that heat capacity is a *unique function* of temperature. The graphs 'a' and 'b' are positioned under one another on the same temperature scale to reveal that the shape and location of the peaks are very close to the *first derivative* of the $m_H(T)$ (dotted curves). It remains to note that the *latent heat* of the phase transition must be proportional to $dm_H/dT$. Thus, the latent heat of the first-order phase transition, lost in the numerous calorimetric studies, was found, simultaneously eliminating the great theoretical mystery.

### 3.4. Limitation of Adiabatic Calorimetry

The Dinichert's work had not changed the interpretation of the λ-peak from "heat capacity" to "latent heat", even though all the attributes of a structural first-order phase transition have there been exposed. A partial explanation of that is: one had to go beyond the abstract term "first order" and its superficial indicators "jump/no jump" or "latent heat/no latent heat", straight to its physical essence, which is a *quantitative* transfer of the matter at interfaces between the competing crystal phases. And that was not the case. But there was also second reason, and it was hidden in the calorimetric technique itself.

The goal of numerous calorimetric studies of λ-peaks in $NH_4Cl$ and other substances was to delineate the shape of these peaks with the greatest possible precision. An adiabatic calorimetry, it seemed, suited best to achieve it. The adiabatic calorimeters, however, are only "one-way" instruments in the sense the measurements can be carried out only as a function of increasing temperature. In the case under consideration, however, it was vital to perform both temperature-ascending and descending runs – otherwise the existence of hysteresis would not be detected. For example, in [16] the transition in $NH_4Cl$ was claimed to occur at fixed temperature $T_\lambda = 245.502 \pm 0.004$ K defined as a position of

the λ-peak. The high precision of measurements was useless: that $T_\lambda$ exceeded $T_o$ by 3°.

The results by Extermann and Weigle were not typical. The kind of calorimetry they utilized permitted both ascending and descending runs. That was a significant advantage over the adiabatic calorimetry used by others in the subsequent years. But there was also a shortcoming both in their and the adiabatic calorimetry techniques resulted in the unnoticed error in the presentation of the λ-peaks in Fig. 2b: the *exothermic latent heat* peak in the descending run had to be *negative* (looking downward). Adiabatic calorimetry, in spite of all its precision, could not detect that.

### 3.5. Straightforward Test of the Heat Effect in NH4Cl

Differential scanning calorimetry (DSC) is free of the above shortcomings [23]. Carrying out temperature descending runs with DSC is as easy as ascending runs. Most importantly, it displays endothermic and exothermic peaks with *opposite* signs in the chart recordings, which results from the manner the signal is measured. If the λ-peak in $NH_4Cl$ is a *latent heat* of phase transition, as was concluded above, the peak in a descending run must be exothermic and look downward. Our strip-chart recordings made with a Perkin-Elmer DSC-1B instrument immediately revealed [13] that the peak acquires opposite sign in the reverse run (Fig. 3). Its hysteresis was also unveiled.

### 3.6. Consequential Mistake: "λ-Anomaly" in Liquid Helium is Latent Heat of Phase Transition, not Heat Capacity

It is not realistic to continue assigning a *heat capacity* to the λ-peak in liquid He, while the same peculiar peak in $NH_4Cl$ is proven to be a *latent heat*. Looking into the experimental techniques utilized in the investigations of the liquid He phase transition we find that, indeed, it is the adiabatic calorimetry with its inherent limitations that was used. The critical claim that latent heat is absent in that transition was made on the basis that it was not observed in passing along the λ-curve. The irony is that the whole λ-peak is none other than a latent heat.

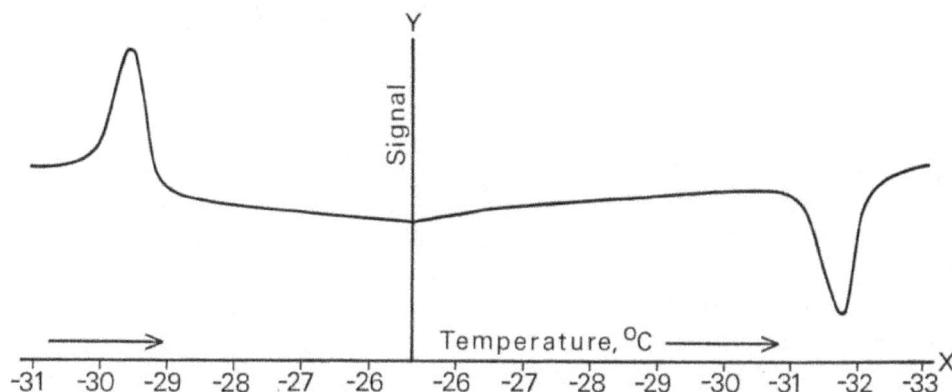

**Figure 3.** The actual DSC recording of NH4Cl phase transition cycle, displaying temperature-ascending and descending peaks as endothermic and exothermic accordingly, thus delivering visual proof of their latent heat nature

The distinction between phase transitions in solid and liquid states is not so wide as it may seem. In liquid *He* we deal with the *liquid polymorphism* – a phenomenon possibly unknown when its λ-peak was discovered, but is known presently [24]. The liquid is not a completely disordered state. At any given moment it consists of multiple tiny clusters of approximately closely packed particles. A certain kind of a *short-range* order within the clusters is preserved, analogous to the *long-range* order of crystals. Change in the manner the particles are packed in the clusters with temperature (or pressure) is similar to polymorphic phase transitions in crystals.

*We maintain that measurements of the heat effect in the liquid He phase transition with an appropriate calorimeter (such as differential scanning calorimeter) will produce the peak in cooling and heating runs as looking in opposite directions (as in Fig. 3), thus exhibiting its latent heat nature. Simultaneously, its hysteresis will be revealed, in which case the λ-peak in the cooling run will be shifted from its current temperature 2.19 K toward 1 K.*

The physical phenomenon used by Ehrenfest to introduce the *second-order* type of phase transitions was not existed. Neither the subsequent voluminous work over many years on shape delineation and analytic description of the "specific heat λ-anomaly" has any scientific value.

# 4. Nucleation-and-Growth Mechanism of Solid-State Phase Transitions

The *nucleation-and-growth* molecular mechanism of solid-state phase transitions was derived by Mnyukh in 1971-1979 from the results of systematic experimental studies [25-38], later summarized in [14] and finalized in [39-45]. Solid-state phase transitions were found to be a *crystal growth,* very similar to the crystal growth from the liquid phase, but involving a distinctive kind of nucleation and a specific *contact* structure of interfaces.

It was demonstrated that the *nucleation-and growth* mechanism is general to all kinds of solid-state phase transitions, including ferromagnetic, ferroelectric and order-disorder. It will be now briefly described, for it covers the *normal – superconducting* phase transitions as well.

## 4.1. Nucleation in Microcavities. Hysteresis

Nucleation of the alternative phase in solid-state reactions differs in all respects from the theoretical-born fluctuation-based statistical process described in the Landau and Lifshitz classical textbook (Section 150 in [6]). Nucleation in a given crystal is pre-determined as to its location and temperature. It would not occur at all in perfect crystals, but real crystals are never perfect. The nucleation sites are located in specific crystal defects – microcavities of a certain optimum size. These defects contain information on the condition (*e.g.,* temperature) of their activation. Nucleation lags are not the same in different defects, but are finite. *Hysteresis is a necessary component if the phase*

*transition mechanism.* The "ideal phase transition" without hysteresis at the temperature $T_o$ ($F_1 = F_2$) cannot occur, considering that the thermodynamic driving force $\Delta F = f(T-T_o) = 0$ at $T = T_o$. Taking into account that the nucleation lags occur in both directions of the phase transition, some finite *hysteresis* of phase transitions is inevitable. Its value, however, is not exactly reproducible.

## 4.2. Structural Rearrangement at Contact Interface. Phase Coexistence

A solid-state phase transition is an intrinsically *local* process. It proceeds by "molecule-by-molecule" structural rearrangement at interfaces only, while the rest bulk of the original and the emerged phases remain static (Fig.4). This mechanism is in compliance with the requirement *'A'* in Section 2, namely, no macroscopic "jumps" occur during the phase transition. The seeming "jumps" are simply the differences between physical properties of the initial and resultant phases, appearing in the experiments as "jumps" when the transition range is either narrow enough, or passed quickly, or both.

The coexistence of the phases during phase transition is self-evident.

## 4.3. Epitaxial Phase Transitions

**Figure 4.** Model of a contact interface. Phase transition proceeds by molecule-by-molecule building up new layers of the emerging (upper part) phase. It is in accord with the thermodynamic condition 'A' that requires emerging only an "infinitesimal" quantity of the new phase at a time

Fig. 4 illustrates a case when crystal orientations of the phases are not related. This is because the nuclei that initiate phase transitions in crystal defects have, in principle, arbitrary orientations. However, a structural *orientation relationship* (OR) is frequently observed. That does not mean these transitions occur by a "deformation" or "distortion", involving "displacement" of all molecules, as still overwhelmingly believed. They materialize by nucleation and growth as well, no matter how minute the seeming "distortion" could be in some cases.

The OR can be rigorous or not. There are two circumstances when the OR is rigorous. One is in pronounced layered crystal structures. A layered crystal consists of strongly bound, energetically advantageous two-dimensional units − molecular layers − usually appearing almost unchanged in both phases. The interlayer interaction in these crystals is relatively weak by definition. The difference in the total free energies of the two structural variants is small. (This is why layered crystals are prone to polymorphism). The phase transition is mainly resulted in changing of the mode of layer stacking, while layers of the new phase retain the previous direction. Nucleation occurs in the tiny interlayer cracks. Given the close structural similarity of the layers in the two polymorphs, *this nucleation is epitaxial* (oriented) due to the orienting effect of the substrate (opposite surface of the crack). Another case of the rigorous *epitaxial nucleation* is when the unit cell parameters of the polymorphs are extremely close (roughly less than 1%) even in non-layered crystals, as in the case of ferromagnetic phase transition of iron (Section 4.2 in [14]).

*Epitaxial* phase transitions exhibit themselves in a specific way: the OR is preserved, the x-ray Laue-patterns of the phases appear almost identical, hysteresis can be small and easily missed, so is latent heat, etc. Without a scrupulous verification, the phase transitions may be taken for "instantaneous", "cooperative", "second-order", *etc*. In the case of the 1924 x-ray study of lead [46] it was the cause of the erroneous conclusion on the structural identity of the pre-superconducting and superconducting phases.

### 4.4. Identification of the Phase Transition Order

The Landau's definition of seconds-order phase transitions provides distinctive details to the option *'B'* in Section 2. Any deviation from its strict conditions, however small, would violate their physical meaning. In each such case the transition would be not a second order, but first, that is, materialized by a reconstruction at interfaces rather than homogeneously in the bulk. Identification of a second-order phase transition must not be "approximate"; in this respect it should be noted that its "pure" example has never been found. The core of the definition of second-order phase transition is the absence of *any* hysteresis; this alone had to prevent the normal- superconducting phase transitions to be classified as such, for all of them exhibit hysteresis, large or small.

In order to distinguish between the two kinds of transition a reliable indicator must be used to tell whether the process is localized at interfaces, where transfer of the material between the coexisting phases takes place (first order), or it is a qualitative change involving the whole bulk simultaneously (second order). The reliable indicators of *first-order* phase transitions are *interface, heterophase state, hysteresis of any physical property, latent heat*. Any one is sufficient, for all four are intimately connected. Any one of them will guarantee all being present. Detection of a two-phase coexistence in any proportion at any temperature would prove the transition being first order.

Thus, identification of a first-order transition is simple and definite. Not so with second-order phase transitions. Physical state of the two phases at the "critical point $T_c$" must be identical. Proving a case requires to show that all the above first-order indicators are absent. Some can be overlooked or remain beyond the instrumental capability. The same is true for "jumps": their seeming absence cannot serve as a reliable indicator as well. It is the small or undetected "jumps" that were frequent source of erroneous classification. The least reliable but frequently used indicator of second-order phase transition is absence of latent heat. Not only the latent heat can be too small for detection, it is prone to be confused with a heat capacity even when observed, as demonstrated above.

## 5. Normal – Superconducting Phase Transitions: Common Treatment

### 5.1. Assigning the "Order"

Some phase transitions, both the normal-superconducting and otherwise, are still being frequently called "second-order" in the literature. When assigning "second order" to a phase transition, it is a common practice to use only a single rational while disregarding evidence unequivocally indicating the first-order transition. It is always a blunt statement lacking sufficient argumentation. Different criteria are used. In superconductors it is the seeming crystal-structure phase identity. Or it can be latent heat as, for example, the claim [47-49] that a normal- to superconducting transition changes from first order (with a latent heat) to second order (without latent heat) with increasing magnetic field. Sometimes it is a missed (or "too small") "jump" of some property. Sometimes it is the narrow range of transition designated to be a "critical point $T_c$". Finally, if it is possible to be more wrong than all the above, it is the "specific heat lambda-anomaly", because it is not a specific heat, but the latent heat of polymorphic transition.

Proper verification of the remaining "second order" phase transitions will turn them to first order. In fact, a slow, but steady process of second-to-first-order reclassification is going on. The ferromagnetic phase transition in Fe at ~769°C is a glowing example. For decades, it was regarded as best representative of second-order phase transitions. The two involved crystal phases, just like in superconductors, were believed to be identical and even marked as a single crystal phase. No "jumps" were ever reported. But it was shown (Sections 4.2 and 4.7 in [14]) that a change of the crystal structure in that phase transition had been overlooked. What's more, it was explained [40] that *all* ferromagnetic phase transitions are "magnetostructural". The latent heat of the ferromagnetic phase transition in Fe was ultimately recorded [50].

### 5.2. "Weakly First-Order" Phase Transitions

The "order" problem of normal – superconducting phase transitions is still unsettled in the current literature. It became even more cumbersome due to the better awareness that they do not fit second order as defined, considering that they occur over a temperature interval rather than in a fixed point and exhibit hysteresis.

Phenomenologically, they matched to the first order better. Without sacrificing scientific purity they hardly could be regarded a "critical phenomenon" and, as such, suitable for treatment by statistical mechanics. Yet, the out-of-date claims that these phase transitions are second order on the ground that there is no latent heat (which is factually incorrect) are not extinct completely. A single answer to whether all normal – superconducting phase transitions are a first or second order is not existing anymore: some are now called first order, others are called second order. Moreover, as already noted, there are claims that one and the same phase transition can have different "order" under different conditions.

Trying to salvage the area of application of statistical mechanics, the theorists resorted to a compromise – "*weakly first order*" phase transitions – and treated them as *second order*. No attempt is made to look into the irreconcilable physical difference between the molecular mechanisms of the first- and second-order phase transitions. Instead, the approach is entirely phenomenological. As such the two mutually exclusive processes became reconcilable. Definitive information about the "weakly first-order phase transitions" is absent. In general, the effects accompanying phase transition – change of crystal structure, transition range, energy jumps, latent heat, discontinuities in specific volume, hysteresis – should be "small". How small? It depends on the researcher's perception. If the decision is made that a first-order phase transition is "weakly first order", then its treatment as a *continuous* is considered permissible.

Application of statistical mechanics assumes that the process under investigation is based on a fluctuation dynamics involving all particles in the bulk simultaneously. But the physical process in the first-order phase transitions is opposite, no matter how "weakly" they are. They materialize by restructuring that takes place only at the interface between the two contacting crystal phases. The process involves only a single particle at a time, while the crystal bulk on the both sides of the interface remains static (Fig. 4). Obviously, the application of statistical mechanics to the non-statistical process is quite inappropriate.

The following simple rule will eliminate all confusion regarding the order of any phase transition in the solid state. If the process is homogeneous in the bulk, then it is *second* order, and if it is localized on interfaces, it is *first* order. We can assure the readers that the latter will always be the case – even in the "weakliest" first-order phase transitions.

### 5.3. "Anomalies" at Tc: Distortions, Displacements, Deformations

Since 1924 the *concept of structural identity* (CSI for brevity) of normal and superconducting phases is impairing the investigation of superconductivity. It is instructive to trace it chronologically.

- (1924) The CSI, inferred from the investigation of a single superconductor (lead) [46] was unconditionally accepted by everyone as valid for all superconductors,

thus unwittingly impairing the ensuing scientific research.

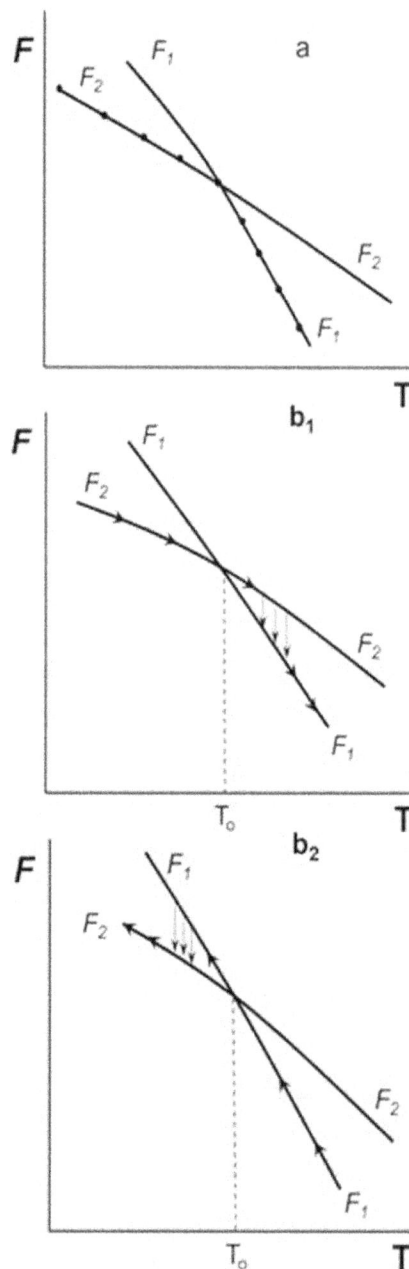

**Figure 5.**   a. Two curves of free energy, $F_1$, and $F_2$, representing two phases involved in a first-order phase transition. The dotted portion of each line indicates the thermodynamically stable phase. The scheme does not exactly reflect the actual behavior of the real system. Phase transition will not proceed without lowering the free energy of the system and, therefore, will not occur at the crossing point where $F_1 = F_2$.

**b1, b2**. Behavior of a real system when hysteresis provides a driving force for phase transition to occur. Range of transition is also schematically taken into account. **b1** heating, **b2** cooling

- (1934) Rutgers [51] published his work, applying Ehrenfest's second-order phase transitions to superconductors. The CSI played a major role in that endeavor. As we presently know, the phase transitions in question are not second order.

- (1952) The CSI is holding firm. Nobody still expressed doubts that it might be incorrect or, at least, not covering all superconductors. It diminished the value of Laue work on superconductivity [52], especially his thermodynamic theory of normal-superconducting phase transitions where the CSI was quite essential.

- (1972) Publication of three articles by H. R. Ott [53-55] where it was revealed that single crystals of lead, aluminum, zinc and gallium undergo dimensional and volume changes upon their normal-superconducting phase transitions. That work ruined the CSI by proving that the normal and superconducting phases are different crystal structures.

- (1987) Evidence of "structural distortion" at $T_c$ in some high-temperature superconductors [56, 57].

- (1991) The CSI is alive, however; the Ott work did not have the impact it deserved. P. F. Dahl [1] still describes CSI as decisively proven. There is no indication that anyone claims otherwise.

- (1992) In the frenzy of the experimental investigations of the "high-temperature" superconductors after their discovery in 1986, the CSI may seem to silently left behind. The question whether CSI is right or wrong is not raised in the experimental literature. Instead, experimental studies were aimed at finding what changes *in* the crystal structure occur at the transition into superconducting state. A number of those studies, performed on different superconductors, in which several highly sensitive experimental techniques were employed, are collected in the book "*Lattice Effects in High-$T_c$ Superconductors*" [58]. Ample evidence is presented that *there are* certain changes at $T_c$. These changes are reported as lattice *distortions*, thermal expansion *jumps*, atomic *displacements* from regular lattice sites, local *displacements* from average crystal structure, *changes* of some inter-atomic distances, etc.

- (1992-2015) The reports on a "structure anomaly", "lattice distortion", or alike, at $T_c$ in different high-temperature superconductors continued to appear, e. g. [59-62], suggesting that the phenomenon is general to all superconductors. Treatment in the literature of all those distortions, jumps and displacements at $T_c$ has a general trait: the superconducting crystal structure is actually regarded as the same pre-superconducting crystal structure, only somewhat modified. Even though the CSI is not openly present there, its shadow invisibly affects the scientific thinking. The following question could (but did not) arise: why is it the superconducting phase that becomes distorted? Would not be it also fair to regard the pre-superconducting phase becoming distorted in case of a superconducting-to-normal phase transition? There is no need to answer. Neither of the two phases is distorted. The phase transitions in superconductors and non-superconductors alike do not occur by displacements, distortions or deformations. They proceed in both directions by a nucleation-and-growth. Notwithstanding how minute the structural changes could be, the phase transition is a *replacement*, and not *modification*, of the old structure. The two crystal structures, no matter how similar, are built according to the rules of crystallography and minimum free energy.

- (2015) Disregarding the Ott's experimental results and the numerous reports on structural "distortions" at $T_c$, the CSI is still holding in the minds of some theorists. Recent examples: (1) "High-resolution X-ray data show no change in crystal structure at $T_c$, indicating no first-order transition" [63]. (2) "Measurements show that at the superconducting transition there are no changes in the crystal structure or the latent heat release and similar phenomena characteristic of first-order transitions". "The complete absence of changes of the crystal lattice structure, proven by X-ray measurements, suggests that ..." [64]. (3) "When a superconductor is cooled below its critical temperature, its electronic properties are altered appreciably, but no change in the crystal structure is revealed by X-ray crystallographic studies" [65].

# 6. Justy and Laue Were Correct, but Ignored and Forgotten

## 6.1. Why Second-Order Phase Transitions are Impossible

Second-order phase transitions were theoretically created by Ehrenfest [2] as follows. He used the thermodynamic free energy $F$ as a function of other thermodynamic variables. Phase transitions were classified by the lowest $F$ derivative which is discontinuous at the transition. *First-order* phase transitions reveal a discontinuity in the first $F(T,p)$ derivative (temperature T, pressure $p$). *Second- order* phase transitions are continuous in the first derivative, but show discontinuity in the second derivative of $F(T, p)$: there is no latent heat, no entropy change, no volume change, and no phase coexistence.

Justi and Laue in their lecture (delivered by Laue) at the September 1934 meeting of the German Physical Society in Bad Pyrmont [66] rejected the possibility of second-order phase transition. The presentation was of a general significance, not limited by superconductors. Laue made strong thermodynamical arguments against the physical realization of Ehrenfest's criteria for second-order phase transitions. He analyzed $F(T)$ for two competing phases. In case of a first-order phase transition the two $F(T)$ curves intersect (Fig. 5). In case of a second-order transition these curves do not intersect, but osculate at the alleged critical point and then either separate again (Fig. **6a**), or merge (Fig. **6b**). In both cases the phase represented by the lower curve remains stable in all temperatures, unchallenged at the point of osculation. It was a proof that *second-order phase transitions cannot materialize*.

There was only a single opposition to the Justi and Laue analysis. It surfaced much later and was marked by a dishonesty. Thirty years after the meeting in Bad Pyrmont, Gorter [67] published a review with recollections of Justi and Laue presentation. He stated that he attended this meeting and pointed out to Laue that his diagrams are incorrect. He wrote: "With Laue, Keesom and I had rather tenacious discussions". This is not true. The published discussion of the presentation reveals that Gorter only suggested that order of phase transition may not necessarily be an integer number. This comment, let alone being unsound, was not significant for the essence of the Justi and Laue presentation and could hardly be called a "tenacious discussion". Then Gorter went on saying in the review that by introducing an "internal parameter indicating a degree of superconductivity" he and Casimir demonstrated in a short communication that the Justi and Laue osculating diagram is incorrect. Indeed, Gorter and Casimir have presented their short-lived hypothesis on "degree of superconductivity" by publishing the same article in two different journals and a book chapter [68-70].

However, there was no critique of Justi and Laue presentation, nor the osculating diagram. No wonder Laue never responded to the Gorter's critique, since there was no one. Gorter completed his recollection on the topic by saying "It is clear that in this model there is no place for Laue and Justi's objection". We have two notes to that statement. (1) The reverse is true: there is no place for a theoretical model that is not in compliance with thermodynamics. (2) Calling the proof by Justi and Laue an "objection" reduced the result of the rigorous thermodynamic analysis to the rank of an opinion. Laue did not change his position in the later published book and the article on superconductivity [52, 71].

The Gorter review was published four years after Laue death, so Laue could not already respond to it.

*Justi and Laue must be credited for finding that second-order phase transitions cannot exist. Unfortunately, their critically important contribution was damaged by Gorter by providing a pretext to ignore it, as did L. Landau, creator of the theory of second-order phase transitions.*

### 6.2. Laue's Further Efforts

At the time when Justy and Laue tried to persuade the contemporaries that second-order phase transitions cannot exist, the *nucleation-and-growth* mechanism of solid-state phase transitions, as described in Section 5, was not yet discovered. In spite of their efforts, a general belief has been firmly established that second-order phase transitions are real, while the normal – superconducting transitions were their primary case. In that environment Laue in his 1952 book [52] concentrated narrowly on *normal – superconducting* phase transitions. Not using the term "second order", he did his best to show that they are not.

To make a strong case a positive description of non-second-order phase transitions in superconductors had to be given. There was a major hurdle in the way: the "structural identity" of the phases. Like everyone else, Laue erroneously accepted it as a fact. We read in his book (p. 4): "The transition from normal to superconductor does not change the form or the volume of the specimen; its lattice remains the same not only in is symmetry but also in its three lattice constants. This was proved for the lead by Kamerlingh-Onnes and Keesom using X-ray analysis". It had taken another 20 years before it was shown that the form and volume of single-crystal superconductors do change.

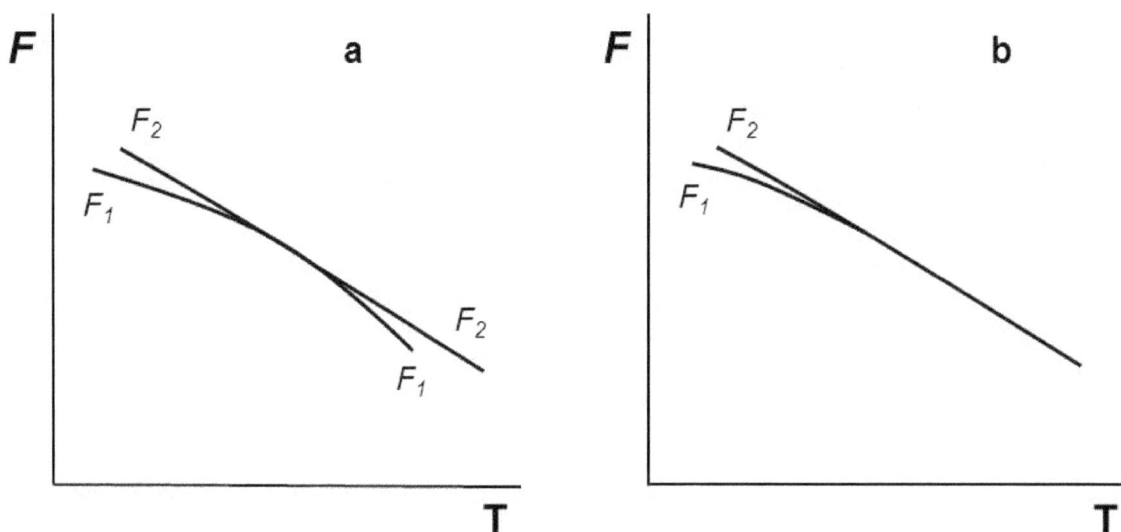

**Figure 6.**  Pairs of curves of free energy, $F_1$ and $F_2$, involved in the alleged second-order phase transition: **a** in the case of touching; **b** in the case of merging

In the absence of a consistent picture of "non-second order" phase transitions, Laue could only demonstrate that *normal – superconducting* phase transitions are incompatible with being second order. He argued: "The [normal – superconducting] transition is by no means continuous", and "In the present case one solid phase changes directly into the other, which is in contact with it", and "The 'intermediate state' almost always appears, a mechanical mixture of normal and superconducting parts", and "The transition from the superconducting to the normally conducting state requires heat: the converse process liberates heat". These properties were quite sufficient to reject its second-order nature. In retrospect, we deal here with the inalienable attributes of the nucleation-and-growth mechanism of solid-state phase transitions. But that was not enough for others: for them a normal-superconducting phase transition with its "structural identity" still looked like "second order". The misconception survived in some places to the present days. A question whether phase transitions in superconductors are the first or second order is not being settled definitely. Even worth: they incresingly seemed compatible with neither. But we eliminate this problem now: they all have the nucleation-and-growth mechanism.

In his interpretation of *normal – superconducting* phase transitions Laue was way ahead of his contemporaries, but still far from understanding the phenomenon. For example, phase transitions by "jumps" were not rejected, the word "nucleation" was not even present in the considerations, the hysteresis was incorrectly ascribed to relaxation effects, the cause of the transition range and uncertainty in the location of transition temperature remained unknown, *etc.*

### 6.3. Forty Years Later: They are not Found

Since 1934, the existence of second-order phase transitions became an unbreakable dogma. Forty years later, in spite of the unfavorable environment, claims of their non-existence were resumed (Mnyukh [30, 14, 41, 44, 36]), growing more categorical with years.

(1973) "Polymorphic transitions of second- or 'higher' order are not found"; "As for the different classifications of the phase transitions in solids, … they will, most probably, simply be reduced to polymorphic transitions of the *epitaxial* or *non-epitaxial* type".

(1978) "It is now impossible to find even one well-documented example of a second-order polymorphic transition!"

(2001) "Only one problem with the classification by first and second order will remain, namely, to find at least

one well-proven crystal phase transition that would be not of the first order".

(2013) "True second-order phase transitions will never be found. The first / second-order classification is destined to be laid to rest."

(2014) "Essential result of [our] studies was the conclusion that second-order phase transitions do not exist".

These "heretical" statements resulted from a comprehensive study of solid-state phase transitions undertaken from 1960th. They led to the discovery of the *nucleation-and-growth* molecular mechanism of polymorphic phase transitions (outlined in Sections 4.1, 4.2), universal to all phase transitions called "structural" in the literature. It was followed by the discovery of its epitaxial variation (Section 4.3) which is usually mistaken for "second order" phase transitions. No reasonable niche to accommodate second-order phase transitions remained.

A major milestone was the establishment of a structural nature of *ferromagnetic* phase transitions, considering that they have been used in the literature as the last resort for second-order phase transitions. It was shown [40] that all ferromagnetic phase transitions must materialize by a crystal rearrangement at interfaces by the *epitaxial* mechanism. The same is true for ferroelectric phase transitions. Adding to the above are *order-disorder* phase transitions. They were demonstrated to proceed also by a nucleation and growth of the orientation-disordered crystals in the original crystal medium [32]. The "heat capacity λ-anomaltes" almost completed the picture. They used to be the reason for the introduction of the second-order type of phase transitions (see Section 3.1), but turned out to be a *latent heat* of the nucleation-and-growth.

Only the phase transitions in superconductors were not directly analyzed previously. Not any more: we have shown in this article that they fall in line with all other solid-state phase transitions. There is enough evidence that second-order phase transitions do not exist at all, being a product of excessive theoretical creativity.

## 7. Example

This section is to illustrate how familiarity with the molecular mechanism of solid-state phase transitions would eliminate serious misinterpretations and wrong conclusions detrimental to the science of superconductivity.

In the considered work [49] the specific heat of a high-quality single crystal of superconductor $ZrB_{12}$ was measured with and without an applied magnetic field. Application of magnetic field of increasing strength shifted the temperature of normal – superconducting phase transition (~ 6 K in zero field) down. The recordings of specific heat $C = f(T)$ were typical "λ-anomalies". The recorded value was a combined contribution of specific heat $C$ and latent heat $Q$, even though only $C$ was shown on the ordinate axis. Without magnetic field the λ-peak was high and thin and would traditionally be identified with second-order transitions. This time, however, it was assigned first order. As the applied magnetic field was increased, the λ-peaks became lower and wider until degrading into a diffuse hump at ~ 1 K. This hump was assigned to represent a second-order phase transition. It was concluded that *"the normal-to-superconducting state transition changes from the first order (with a latent heat) to second order (without latent heat) with increasing magnetic field"*. Nothing was said about the non-integer "order" of the intermediate cases between these two extremes. *Hysteresis* and *range of transition* were present in all recordings, but had no role in that claim. In fact, they were enough to invalidate the presence of a second-order transition, but the only used criterion for assigning the "order" was latent heat.

But the most revealing misinterpretation that nullifies the theoretical claims of that work, including about a type of the superconductor, was in the wrong identification of the *latent heat* and *the specific heat* in the experiments. It is due to this mix-up the improbable phenomenon of turning first-order phase transition into second order (and vice versa) entered the scientific literature. According to the authors, "the latent heat [is] the area below the specific-heat peaks". *But it is the λ-peaks that are a latent heat*. The latent heat peaks rest, as on a baseline, on the curve delineating specific heat over the range of phase transition (Fig. 7).

## 8. Structural Approach to Phase Transitions in Superconductors

Nonexistence of second-order phase transitions means that all normal – superconducting phase transitions are first order. Considering that being first order is equivalent to materialize by changing the crystal structure, and the change has the *nucleation-and-growth* mechanism, we have arrived at the following conclusions.

▶ The crystal structures of the normal and superconducting phase are not identical. The inference from the x-ray data made in the early years of superconductivity that there was no change of the crystal structure at $T_c$ was in error, probably resulted from the experimental imperfections and the *epitaxial* type of the investigated phase transition when the difference is difficult to detect.

▶ The phase transitions in question should be approached as being primarily *structural*. It is important to correctly identify the cause and the effect. The cause in every solid-state phase transition is crystal rearrangement, and the effect is the physical properties of the new crystal structure. In our case the new property of the resultant crystal is its superconducting state. In order that a phase transition could occur, two crystal versions of almost equal free energies $F_1 \approx F_2$ must exist. The towering component in the both $F_1$ and $F_2$ is the energy of chemical bonding. The contribution due to superconductivity can only affect the balance between $F_1$ and $F_2$ in favor of one or the other, but not to cause a phase transition if the $F_1 \approx F_2$ does not exist. Therefore, it is the structural phase transition that gives rise to superconductivity, and not *vice versa*. *Superconductivity is a property of a specific crystal structure.*

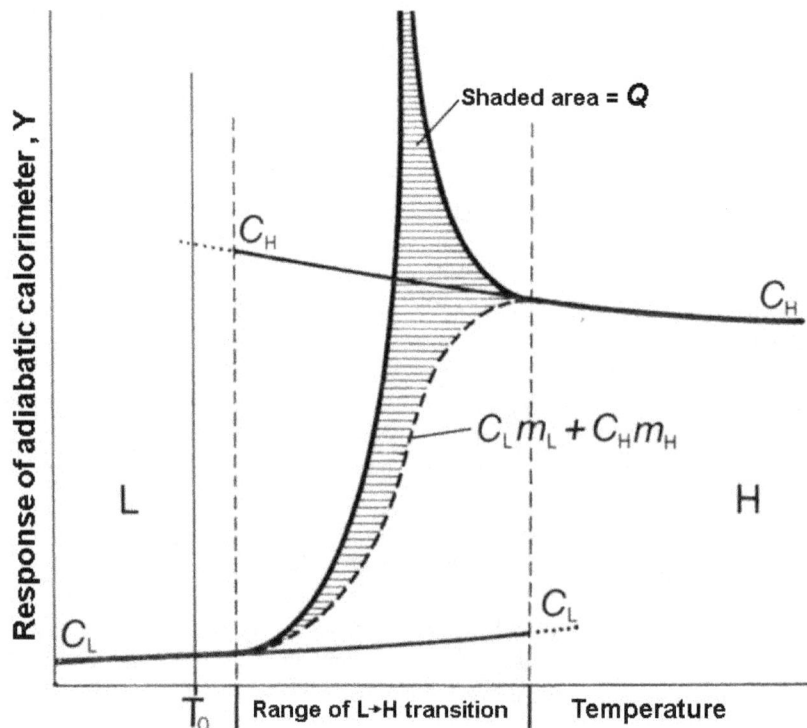

**Figure 7.**  Heat capacity $C$ and latent heat $Q$ in the range of transition between phases marked L and H. The drawing features: Two separate overlapping plots $C_L(T)$ and $C_H(T)$ would represent the specific heat of each phase. The S-curve (dashed bold) represents an amalgamated plot of two independent contributions $C_L$ and $C_H$ from the coexisting phases L and H, taking into account their mass fractions $m_L$ and $m_H$; $m_L + m_H = 1$. The bold-faced curve with "λ-anomaly", which would be produced by calorimetry. It is composed of both $C_L$ and $C_H$ contributions (dashed curve) and a superstructure of the latent heat $Q$ of the phase transition. (Reproduced from [14])

▶ In view of the fact that a *normal – superconducting* phase transition has the same universal *nucleation-and-growth* molecular mechanism as all other solid-state phase transitions, it bears no relation to the physical nature of superconductivity. That does not mean a *comparison* of the pre- superconducting and superconducting structures is useless. Quite opposite: it could bring valuable information, especially in the epitaxial cases. Indeed: why does such a minor crystal-structure change produce that drastic change in electrical conductivity? Because of the belief in the structural identity of the phases, this direction of research had been excluded up to the time of discovery of the high- temperature superconductivity in 1986, and is still negatively affecting scientific thought.

▶ The phenomena believed to be a deviation from the "ideal" equilibrium *normal – superconducting* phase transition – range of transition, hysteresis, phase coexistence, uncertain transition temperature – receive now final explanation. They are simply the inalienable properties of the universal *nucleation-and-growth* mechanism, which is the only way solid-state phase transitions materialize.

▶ The problem of exact temperature of normal – superconducting phase transition (and any other phase transition for that matter) is also clarified. It is not the temperature at which the resistance is one half of the value just before the drop, as it is commonly defined. Neither it is located at the foot of that curve, as Laue suggested. From a thermodynamic point of view, it should be the position of $T_o$ when $F_1 = F_2$, but it is not directly achievable in experiments, considering that any measured temperature will be either above or below the $T_o$ due to hysteresis. The proper (but still approximate) position of the $T_o$ can be found only by extrapolation of the measured temperatures in heating and cooling runs. The $T_o$ will be somewhere in between; it is the only constant temperature characteristic of that particular phase transition.

Calling the *observed* temperature "critical temperature (or point) $T_c$" is incorrect: crystal growth is not a "critical" phenomenon eligible to be treated by statistical mechanics. Neither is it a fixed point. Due to the hysteresis it is even possible to have superconductivity at room temperature. The only experimental problem would be to grow sufficiently defect-free single crystal at the temperature when it is still in the superconducting state. The superconducting state at $T_{room}$ will not be stable, however.

▶ The foundations of the Ginzburg-Landau theory of superconductivity are questioned. (1) It erroneously implies that the mechanism of a *normal – superconducting* phase transition is specific to that type of phase transitions. However, it is general to all solid-state reactions and, therefore, has nothing to do with the resulting superconductivity. (2) The theory in question is based on the Landau theory of seconds-order phase transitions. Not only they are nonexistent, their alleged molecular mechanism is antithetical to that of real phase transitions.

▶ Realization that normal → superconducting phase transitions occur by a reconstruction of the crystal structure, and that it has the *nucleation- and-growth* mechanism, opens an opportunity to correlate recording of electric conductivity with the physical process in the investigating object. Because high-T superconductors typically have a layered structure, we illustrate it with a layered single crystal. Fig. 8 shows schematically the process of crystal rearrangement during a phase transition in layered single crystals of hexamethyl benzene [36]. That it is reproduced from a work not related to superconductivity is of no significance.

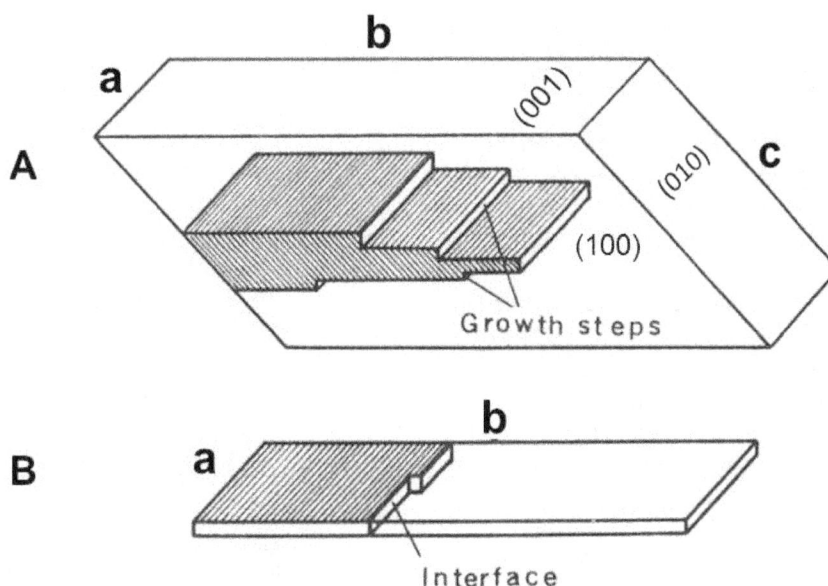

**Figure 8.** **A**. Schematic illustration of phase transition in a layered crystal. It proceeds by growth of the wedge-like crystals of the alternative phase within the initial single crystal. This is an epitaxial type of phase transition – when the molecular layers in the two phases are very similar and retain their orientation in the transition. **B**. Every step of the growing new crystal phase propagates by shuttling movements of kinks along the step

The single-crystal shown in 'A' is brittle, easily cleaved into thin plates parallel to *(001)*. If it is of good quality, the new phase (now we assume it to be superconducting) is developing by epitaxial growth of a wedge-like crystal within it with its flat faces parallel to the *(001)* cleavage planes. The growth steps shown in 'A' move over one another toward *(010)* face. Their movement is by kinks (shown in 'B') shuttling between the *(100)* faces. In the absence of this picture, the researcher will not be able to explain why steepness and shape of the electrical resistivity curve $\rho = f(T)$ depend more on the position of the measuring electrodes than on the physics of the phenomenon. Indeed, if the sample is confined between the electrodes attached to the *(100)* faces, the resistivity drops to zero as soon as the very first shuttling kink makes a connecting string. The $\rho$ drop could be so steep, that its difference from $90^0$ might not be even measurable. At this point, while the $\rho = f(T)$ curve shows the normal $\rightarrow$ superconducting transition already completed, it is barely started, considering the almost whole crystal is still in non-superconducting phase. But some steep before $\rho$ becomes zero can be expected for the measuring electrodes attached to the *(010)* faces, and even more so for the electrodes attached to the *(001)* faces. In the latter case, the $\rho = 0$ is reached only when the almost whole frame of the original sample is filled by superconducting phase. This example may serve as a warning not to assign too much value to the delineation of the $\rho = f(T)$ curves.

## 9. Conclusions

1. A comprehensive evidence is collected, both theoretical and experimental, that second-order phase transitions are non-existent in nature and even unsuited to approximate the real phase transitions: (a) The "λ-anomaly" in liquid *He* is not *a heat capacity*, but *a latent heat of the phase transition*; (b) The thermodynamic proof by Justy and Laue that second-order phase transitions cannot materialize is valid and true today; (c) The *epitaxial* phase transitions are misinterpreted as the second order; (d) Not a single well-documented second-order phase transition in solids exists.

2. Since all solid-state phase transitions are first order, the first/second order classification is not needed.

3. The first-order phase transitions constitute an *overall replacement* of the crystal structure, notwithstanding how minor the change could be. They are not a distortion/deformation of the original structure, or displacements of its certain atoms. It follows that, contrary to the common belief, superconducting phase has an individual crystal structure not identical to that before the transition.

4. Transitions between normal and superconducting crystal phases materialize by the general *nucleation-and-growth* mechanism of all structural rearrangements. It is not specific to phase transitions in

superconductors and, in spite of the widespread belief in the opposite, it sheds no light on physics of superconducting state.

5. A comparison of pre-superconducting and superconducting crystal structures can be highly informative, especially when the structural differences are minute. It may require, however, the detailed structure analysis of the two phases with even higher resolution than presently available.

6. The Ginzburg-Landau theory of superconductivity contains a mysterious aspect. The theory was built on the assumption that mechanism of a normal $\rightarrow$ superconducting phase transition is uniquely related to superconducting state, which is not the case, for it is the same as in all other solid-state reactions. Besides, the theory assumes the transition to be second-order, which is not the case either.

7. The previously unexplained phenomena always observed in the normal − superconducting phase transitions − range of transition, phase coexistence, hysteresis, uncertainty in the position of transition temperature − are the unalienable features of the general *nucleation-and-growth* mechanism of phase transitions.

8. The clarifications provided in the present article should have positive impact on the ongoing research of superconductivity.

## REFERENCES

[1]   Dahl, P.F.: Superconductivity: Its Historical Roots and Development from Mercury to the Ceramic Oxides. American Institute of Physics, New York (1992).

[2]   Ehrenfest, P.: Phasenumwandlungen im ueblichen und erweiterten Sinn, classifiziert nach dem entsprechenden Singularitaeten des thermodynamischen Potentiales. Verhandlingen der Koninklijke Akademie van Wetenschappen. Amsterdam/ 36, 153-157 (1933).

[3]   Megaw, H.: Crystal Structures: A Working Approach. Saunders London (1973).

[4]   Azbel, M.: Preface. In: In: Brout, R. Phase Transitions (Russian ed.). Mir, Moscow (1967).

[5]   Landau, L.D.: Collected Papers of L.D. Landau. Gordan and Breach, New York (1965).

[6]   Landau, L.D., Lifshitz, E.M.: Statistical Physics. Addison-Wesley New York (1969).

[7]   Keesom, W.H., Clusius, K.: Specific heat of liquid helium. Communications from the Physical Laboratory of the University of Leiden 219e, 42-58 (1932).

[8]   Jaeger, G.: The Ehrenfest Classification of Phase Transitions: Introduction and Evolution. Arch Hist Exact Sc. 53(1), 51-81 (1998). doi:10.1007/s004070050021.

[9]   Simon, F.: Analysis of specific heat capacity in low

temperatures. Annalen der Physik 68, 241-280 (1922).

[10] Parsonage, N., Staveley, L.: Disorder in Crystals. Clarendon Press, Oxford (1978).

[11] Anderson, P.: The 1982 Nobel Prize in Physiology or Medicine. Science 218(4574), 763-764 (1982).

[12] Feynman, R., Leighton, R., Sands, M.: The Feynman Lectures on Physics, vol. 2. Addison-Wesley, New York (1964).

[13] Mnyukh, Y.: The cause of lambda-anomalies in $NH_4Cl$. Speculations in Science and Technology 6(3), 275-285 (1983).

[14] Mnyukh, Y.: Fundamentals of Solid-State Phase Transitions, Ferromagnetism and Ferroelectricity, 2d edition, 2d ed. Authorhouse, (2010).

[15] Callanan, J., Weir, R., Staveley, L.: The thermodynamics of mixed crystals of ammonium chloride. I. The heat capacity from 8 K to 476 K of an approximately equimolar mixture. Proceedings of the Royal Society of London A 327, 489-496 (1980).

[16] Chihara, H., Nakamura, M., (1972). Heat capacity of ammonium chloride between 8 and 300 K. Bulletin of the Chemical Society of Japan 45, 133-140 (1972).

[17] Extermann, R., Weigle, J.: Anomalie de la chaleur specifique du chlorune d'ammonium. Helvetica Physica Acta 15, 455-461 (1942).

[18] Schwartz, P.: Order-disorder transition in $NH_4Cl$. III. Specific heat. Physical Review B4(3), 920-928 (1971).

[19] Simon, F., von Simson, C., Ruhemann, M.: An investigation on the specific heat at low temperature. The specific heat of the halides of ammonia between -70° and room temperature. Z. physik. Chem. (1927), 129, 339-48. A129, 339- 348 (1927).

[20] Voronel, V., Garber, S.: On the nature of the orientational transition in ammonium chloride. Soviet Physics JETP 25, 970- 977 (1967).

[21] Ziegler, W., Messer, C.: A Modified Heat Conduction Calorimeter. Journal of the American Chemical Society 63(10), 2694-2700 (1941).

[22] Dinichert, P.: La transformation du $NH_4Cl$ observee par la diffraction des rayons X. Helvetica Physica. Acta 15, 462-475 (1942).

[23] MacNaughton, J.L., Mortimer, C.T.: Differential Scanning Calorimetry. Perkin-Elmer Corporation, (1975).

[24] Poole, P., Torde, G., Angell, C., McMillan, P.: Polymorphic Phase Transitions in Liquids and Glasses. Science 275(5298), 322-323 (1997).

[25] Kitaigorodskii, A., Mnyukh, Y., Asadov, Y.: The polymorphic single crystal – single crystal transformation in p-dichlorobenzene. Soviet Physics - Doclady 8, 127-130 (1963).

[26] Kitaigorodskii, A., Mnyukh, Y., Asadov, Y.: Relationships for single crystal growth during polymorphic transformation. Journal of Physics and Chemistry of Solids 26, 463-472 (1965).

[27] Mnyukh, Y.: Laws of phase transformations in a series of normal paraffins. Journal of Physics and Chemistry of Solids 24, 631-640 (1963).

[28] Mnyukh, Y.: Molecular mechanism of polymorphic transitions. Soviet Physics - Doclady 16, 977-981 (1972).

[29] Mnyukh, Y.: Polymorphic transitions in crystals: nucleation. Journal of Crystal Growth 32, 371-377. (1976).

[30] Mnyukh, Y.: Molecular mechanism of polymorphic transitions. Molecular Crystals and Liquid Crystals 52, 163-200 (1979).

[31] Mnyukh, Y.: Polymorphic transitions in crystals: kinetics. Molecular Crystals and Liquid Crystals 52, 201-217 (1979).

[32] Mnyukh, Y., Musaev, N.: Mechanism of polymorphic transition from the crystalline to the rotational state. Soviet Physics - Doclady 13, 630-633 (1969).

[33] Mnyukh, Y., Musaev, N., Kitaigorodskii, A.: Crystal growth in polymorphic transitions in glutaric acid and hexachloroethane. Soviet Physics - Doclady 12, 409-415 (1967).

[34] Mnyukh, Y., Panfilova, N.: Polymorphic transitions in molecular crystals – II. Mechanism of molecular rearrangement at 'contact' interface. Journal of Physics and Chemistry of Solids 34, 159-170 (1973).

[35] Mnyukh, Y., Panfilova, N.: Nucleation in a single crystal. Soviet Physics - Doclady 20, 344-348 (1975).

[36] Mnyukh, Y., Panfilova, N., Petropavlov, N., Uchvatova, N.: Polymorphic transitions in molecular crystals – III. Journal of Physics and Chemistry of Solids 36, 127-144 (1975).

[37] Mnyukh, Y., Petropavlov, N.: Polymorphic transitions in molecular crystals – I. Orientations of lattices and interfaces. Journal of Physics and Chemistry of Solids 33, 2079-2087 (1972).

[38] Mnyukh, Y., Petropavlov, N., Kitaigorodskii, A.: Laminar growth of crystals during polymorphic transformation. Soviet Physics - Doclady 11, 4-7 (1966).

[39] Y, M.: Mnyukh, Y. Phase transitions in layered crystals. h ttp ://ar xiv.o rg /ab s/1105. 4299. (2011).

[40] Mnyukh, Y.: Ferromagnetic state and phase transitions. Amerian Journal of Condensed Matter Physics 2(5), 109-115 (2012).

[41] Mnyukh, Y.: Second-order phase transitions, L. Landau and his successors. American Journal of Condensed Matter Physics 3(2), 25-30 (2013).

[42] Mnyukh, Y.: Mechanism and kinetics of phase transitions and other reactions in solids. American Journal of Condensed Matter Physics 3(4), 89-103 (2013).

[43] Mnyukh, Y.: Hysteresis of solid-state reactions: its cause and manifestations. Amerian Journal of Condensed Matter Physics 3(5), 142-150 (2013).

[44] Mnyukh, Y., 2014, 4(1): On the phase transitions that cannot materialize. American Journal of Condensed Matter Physics 4(1), 1-12 (2014).

[45] Mnyukh, Y.: Paramagnetic state and phase transitions.

Amerian Journal of Condensed Matter Physics 5(2), 56-59 (2015).

[46] Keesom, W.H., Kamerlingh Onnes, H.: On the question of the possibility of a polymorphic change at the point of transistion into the supraconductive state. Communications from the Physical Laboratory of the University of Leiden 174b, 43-45 (1924).

[47] Shopova, D., Uzunov, D.: Phases and phase transitions in spin-triplet ferromagnetic superconductors. arXiv:cond-mat/0404261v1 (2004).

[48] Uzunov, D.I.: Introduction to the Theory of Critical Phenomena: Mean Fields, Fluctuations, and Renormalization. World Scientific, Bulgarian Acad. Sci. (1993).

[49] Wang, Y., Lortz, R., Paderno, Y., Filippov, V., Abe, S., Tutsch, U., Junod, A.: Specific heat and magnetization of a ZrB12 single crystal: Characterization of a type-II/I superconductor. Physical Review B 72(2) (2005). doi:10.1109/PhysRevB.72.024548.

[50] Yang, S., Ren, X., Song, X.: Evidence for first-order nature of the ferromagnetic transition in Ni, Fe, Co, and CoFe(2)O(4). Physical Review B 78(17) (2008). doi:10.1103/PhysRevB.78.174427.

[51] Rutgers, A.J.: Note on Supraconductivity I. Physica (2), 1055–1058 (1934).

[52] Laue, M.v.: The Theory of Superconductivity. Academic Press, New York (1952).

[53] Ott, H.: Anisotropy in the length change of gallium single crystals at their superconducting transition. Solid State Communications 9(24), 2225-2228 (1971).

[54] Ott, H.: The volume change at the superconducting transition of lead and aluminum. Journal of Low Temperature Physics 9(3), 331-343 (1972).

[55] Ott, H.: Anisotropic length changes at the superconducting transition of zinc. Physics Letters 38(2), 83-84 (1972).

[56] Horn, P.M., Keane, D.T., Held, G.A., Jordan-Sweet, J.L., Kaiser, D.L., Holtzberg, F., Rice, T.M.: Orthorhombic Distortion at the Superconducting Transition in YBa2Cu3O7: Evidence for Anisotropic Pairing. Physical Review Letters 59(24), 2772-2775 (1987).

[57] Muralidharan, P.U.: Analysis of the Relation Between the Orthorhombic Distortion (b-a) and the Superconducting Transition Temperature in YBa$_2$Cu$_{3-x}$M$_x$O$_7 \pm \delta$ (M = Mn$^{4+}$ and Cr$^{3+}$). physica status solidi (a) 123(1), K39-K42 (1991). doi:10.1002/pssa.2211230144.

[58] Bar-Yam, Y., Egami, T., Leon, J.M., Bishop, A.R.: Lattice Effects in High-Tc Superconductors. World Scientific, Santa Fe, NM (1992).

[59] Boehmer, A.E., Hardy, F., Wang, L., Wolf, T., Schweiss, P., Meingast, C.: Superconductivity-induced re-entrance of the orthorhombic distortion in Ba1-xKxFe2As2. Nature Communications 6 (2015). doi:10.1038/ncomms8911.

[60] Maeta, H., Ono, F., Kawabata, C., Saini, N.L.: A search for extra lattice distortions associated with the superconductivity and the charge inhomogeneities in the high-Tc La$_{1.85}$Sr$_{0.15}$CuO$_4$ system. Journal of Physics and Chemistry of Solids 65(8–9), 1445-1448 (2004). doi: http://dx.doi.org/10.1016/j.jpcs.2004.01.011

[61] Oyanagi, H., Zhang, C.: Local Lattice Distortion in Superconducting Cuprates Studied by XAS. Journal of Physics: Conference Series 428(1), 012042 (2013).

[62] Zhang, C.J., Oyanagi, H., Sun, Z.H., Kamihara, Y., Hosono, H.: Low-temperature lattice structure anomaly in the LaFeAsO$_{0.93}$F$_{0.07}$ superconductor by x-ray absorption spectroscopy: Evidence for a strong electron-phonon interaction. Physical Review B 78(21) (2008). doi:10.1103/PhysRevB.78.214513.

[63] Rey, C.M., Malozemoff, A.P.: Fundamentals of superconductivity. In: Rey, C. (ed.) Superconductors in the Power Grid. pp. 29-73. Elsevier, Amsterdam (2015).

[64] Vasiliev, B.V., 2015: Superconductivity and Superfluidity. Science Publishing Group, New York (2015).

[65] Ford, P.J., Saunders, G.A.: The Rise of the Superconductors. CRC Press (2004).

[66] Justi, E., von Laue, M.: Phasengleichgewichte dritter Art. Physikalische Zeitschrift 35, 945-953 (1934).

[67] Gorter, C.J.: Superconductivity until 1940. In Leiden and As Seen From There. Reviews of Modern Physics 36, 3-7 (1964).

[68] Gorter, C., Casimir, H.: Zur Thermodynamik des supraleitenden Zustandes. Physikalische Zeitschrift 35, 963-966 (1934).

[69] Gorter, C.J., Casimir, H.: Zur Thermodynamik des Supraleitenden Zustandes. In: Archives du Musée Teyler: Série III, Vol. VIII Fascicule 1. pp. 55-60. Springer Netherlands, Dordrecht (1935).

[70] Gorter, C.J., Casimir, H.B.G.: Zur Thermodynamik des supraleitenden Zustandes. Zeitschrift für technische Physik 12, 539-542 (1934).

[71] v. Laue, M.: Zur Thermodynamik der Supraleitung. Annalen der Physik 424(1-2), 71-84 (1938). doi:10.1002/andp.19384240111.

# Searching for a Critical Phenomenon

Critical phenomena are a part of physical science that deal with phase transitions accompanied by singularities like "critical opalescence" and "$\lambda$-anomalies". The theory of critical phenomena assumes phase transitions are a cooperative process driven by thermal fluctuations and subject to statistical mechanics. Ferromagnetic phase transition is usually used as a typical critical phenomenon to analyze. Many theoretical physicists viewed the $\lambda$-anomalies as the most important unsolved problem in theoretical physics. In this article hard evidence is presented that the actual molecular mechanism of all phase transitions in solids, including ferromagnetic, is antithesis to the models utilized in the theories of critical phenomena. Real phase transitions materialize by rearrangement of crystal structure according to the universal nucleation-and-growth mechanism. It is the crystal rearrangement which alters the electric, magnetic, optical, etc. properties. The process is not cooperative; thermal fluctuations are not involved; statistical mechanics is not applicable. Another part of this article is devoted to the singularities. (1) "$\lambda$-Anomalies". Believing that these peaks are heat capacity is a case of mistaken identity: they are latent heat of structural phase transitions. The same is true about the notorious "heat capacity $\lambda$-anomaly" in the liquid helium phase transition: it is a latent heat as well. (b) "Critical opalescence". The literature for the subject was examined. The opalescence in solid-state phase transitions, observed by different authors, turns out not fluctuation-related. It is a light scattering by nuclei and interfaces of arising new phase. The only type of phase transition that stays somewhat apart from the above-enumerated is the liquid – gas in its critical point. The case was reconsidered. The physical cause of inability to compress gas into liquid is explained. The observed opalescence is a cloud of tiny drops of liquid phase appearing; no fluctuations are involved. The case is not "critical" either.

American Journal of Condensed Matter Physics 2020, 10(1): 1-13
DOI: 10.5923/j.ajcmp.20201001.01

# Searching for a Critical Phenomenon

## Yuri Mnyukh

76 Peggy Lane, Farmington, CT, USA

**Abstract**  Critical phenomena are a part of physical science that deal with phase transitions accompanied by singularities like "critical opalescence" and "λ-anomalies". The theory of critical phenomena assumes phase transitions are a cooperative process driven by thermal fluctuations and subject to statistical mechanics. Ferromagnetic phase transition is usually used as a typical critical phenomenon to analyze. Many theoretical physicists viewed the λ-anomalies as the most important unsolved problem in theoretical physics. In this article hard evidence is presented that the actual molecular mechanism of all phase transitions in solids, including ferromagnetic, is *antithesis* to the models utilized in the theories of critical phenomena. Real phase transitions materialize by rearrangement of crystal structure according to the universal nucleation-and-growth mechanism. It is the crystal rearrangement which alters the electric, magnetic, optical, *etc.* properties. The process is not cooperative; thermal fluctuations are not involved; statistical mechanics is not applicable. Another part of this article is devoted to the singularities. (1) "λ-Anomalies". Believing that these peaks are *heat capacity* is a case of mistaken identity: they are *latent heat* of structural phase transitions. The same is true about the notorious "heat capacity λ-anomaly" in the liquid helium phase transition: it is a latent heat as well. (b) "Critical opalescence". The literature for the subject was examined. The opalescence in solid-state phase transitions, observed by different authors, turns out not fluctuation-related. It is a light scattering by nuclei and interfaces of arising new phase. The only type of phase transition that stays somewhat apart from the above-enumerated is the *liquid – gas* in its critical point. The case was reconsidered. The physical cause of inability to compress gas into liquid is explained. The observed opalescence is a cloud of tiny drops of liquid phase appearing; no fluctuations are involved. The case is not "critical" either.

**Keywords**  Phase transitions, Critical Phenomena, Critical point, λ-Anomalies, Opalescence, Latent heat, Heat capacity, First order, Second-order, Nucleation, Hysteresis, Interfaces, Ferromagnetic, Ferroelectric, Order-disorder, Superconducting

## 1. Critical Point and Critical Phenomenon

### (a) Not a very critical point.

"Critical points" are the heart of the "critical phenomena" concept. According to Webster Dictionary, the word "critical" means tending to find fault; forming a crisis; dangerous or risky; causing anxiety. But in mathematics and physics that meaning of the word seems to be lost: "critical point" simply designates a point at which some change takes place. Thus, in case of the "critical temperature" $T_C$ of *second-order* phase transitions by Landau [1], there is only change in the crystal symmetry, but not of the crystal lattice itself. And in case of a *liquid – gas* phase transition the critical point ($T_C$, $p_C$) in the *temperature-pressure* phase diagram is simply when the last traces of the liquid phase disappear. The "critical" meaning of the word arises if those changes are accompanied by *singularities*, such as "critical opalescence" and "heat capacity λ-anomalies". In the absence of singularities the point loses its "critical" significance.

### (b) The main property of a critical point.

From a theoretical point of view, the notion of a "critical point" appears as a feature of the *cooperative* process treatable by statistical mechanics. Phase transitions assumed to result from fluctuations over the bulk of the matter. In such a case, any premature or delayed phase transition cannot occur. Allowing deviations from that rigorous rule, however small, is nonsensical. Any "critical phenomenon" must occur exactly at that point, and the point should be the same when approached from the opposite sides (zero hysteresis). Even minor deviations are not acceptable. Therefore, the most important property of a "critical point" must be its *positional stability*. This feature provides a reliable test of whether a phase transition belongs to "critical phenomena" or not.

### (c) And the reality is . . .

In solid state no example with a true "critical point" exists. Observations of the ferromagnetic phase transitions in the region of their "critical point $T_C$" (also called "Curie point") is an example. As Belov [2] put it, *"Many important questions connected with the behavior of materials in the*

* Corresponding author:
yuri@mnyukh.com (Yuri Mnyukh)
Published online at http://journal.sapub.org/ajcmp

*region* [of ferromagnetic transition] *remain unsettled or in dispute. ... These include...the actual temperature behavior of the spontaneous magnetization near the Curie point, the causes of the 'smearing out' of the magnetic transition, ... the existence of 'residual' spontaneous magnetization above the Curie temperature... It even remains unsettled what we should take to be the Curie temperature, and how to determine it".*

### (d) Definition of a critical phenomenon.

There are books with words "critical phenomena" in their titles, but the definition of that notion is missing. In evaluating each particular case we need to know what is behind that notion. It seems, it is always related to a phase transition. If so, is it the fact of the transition, or the singularities as such? We assume it is their combination and define it to be *a phase transition driven by cooperative fluctuations and accompanied by singularities.*

## 2. *Liquid – Gas* Critical Point

### 2.1. Why High Pressure Cannot Liquefy Gas at $T > T_C$

A number of definitions of a "critical point" in the reference material mention the *liquid – gas* critical point as an example. Yet, nobody seems to understand why it exists, namely, why sufficiently high pressure will not liquefy gas at $T>T_C$. To tackle the problem, it should be first noted that, logically, the only reason for that fact can be alteration of the participating particles under high temperatures. We start from the question: what makes a crystal state stable? The answer is: the ability of the particles to bond together into a low energy long-range order with short standard inter-molecular distances. It allows presenting a crystal structure as a molecular *close packing* [3] with the neighboring particles interlocked like gears. There are many ways for the molecules to pack together; which one prevails depends on its density: the higher it is, the lower its free energy $F$. The close-packing model is applicable to liquids as well. Molecular structure of a liquid is not quite chaotic. At every particular moment it consists of temporary tiny molecular clusters with molecules arranged in an imperfect "close-range" packing order where they are bonding like in a crystal state.

Liquefying gas by applying pressure is to overcome the effect of molecular thermal vibrations and rotations that kept molecules apart, squeezing them together to engage their chemical bonding. The increasing vibrations with rising temperature cause the molecules to loose their individual shape, enlarging their effective size. Ultimately, the temperature is reached (it will be marked as $T_C$) when their interlocking ability disappears. No matter how strongly they are now squeezed together, they are not engaging in the chemical bonding due to their inability to come close enough (and under a correct angle) to each other. The gas would not be liquefied.

As to the $T_C$, it is hard to see why *critical opalescence*

would, as claimed, accompany it.

### 2.2. The Cause of Its Opalescence

It is always said that phase transitions in the vicinity of their critical points are accompanied by *critical phenomena,* or *singularities,* caused by fluctuations of something. For example, I. Z. Fisher [4] claimed that in the vicinity of the *liquid – gas* critical point a great increase of density fluctuations and the correlations of those fluctuations take place, giving rise to *critical opalescence,* as well as to change in the velocity and absorption of ultrasound, etc. Stanley [5] stated that critical opalescence is caused by vast fluctuations of density, and many other theoretical physicists share that view.

It should be self-evident, however, that critical opalescence would appear when the thermal fluctuations, whatever they are, increase with the *rising* temperature toward its critical point $T_C$. However, the fluctuation-based nature of the opalescence and, therefore, its "critical" status, is undermined by a simple experimental fact: *the opalescence appears only upon decreasing temperature.* But thermal fluctuations would not intensify with decreasing temperature.

Then, what is the nature of the observed opalescence? Answer: the phase transition *gas → liquid* in question starts *after* the temperature $T_0$ of the phase equilibrium (when $F_{liquid} = F_{gas}$) has been passed. It begins from nucleation of the liquid phase from multiple sites, producing a common fog – a system of tiny drops of liquid, which then grow and merge into the liquid phase with the gas/liquid interface. The process is not fluctuation-based. It is a *nucleation-and-growth.*

The observed opalescence and the related absorption/velocity effects are not singularities that allegedly accompany that "critical phenomenon". Neither were any λ-anomalies reported. A phenomenon without singularities is not "critical".

## 3. The Phenomena Commonly Deemed "Critical"

The discovery in 1932 of a "heat capacity λ-anomaly" in liquid helium phase transition (Fig. 1) was a sensational scientific event. It triggered intensive theoretical efforts over the upcoming years, as described by G. Jaeger [6]:

*"The liquid-helium lambda transition became one of the most important cases in the study of critical phenomena – its true (logarithmic) nature was not understood for more than ten years ... The Ehrenfest scheme [to classify phase transitions by first, second, etc. order] was then extended to include such singularities, most notably by A. Brain Pippard in 1957, with widespread acceptance. During the 1960's these logarithmic infinities were the focus of the investigation of "scaling" by Leo Kadanoff, B. Widom and others. By the 1970s, a radically simplified binary classification of phase transitions into "first-order" and*

*"continuous" transitions was increasingly adopted."*

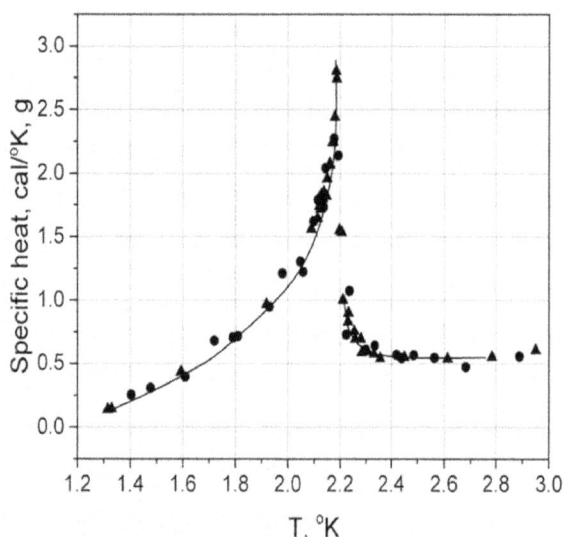

**Figure 1.** The "λ-Anomaly" from calorimetric measurements of liquid helium phase transition. It has been interpreted as a singularity of heat capacity

The attempts to theoretically describe that "critical phenomenon" were continued. In 1982 a Nobel Prize was given to K. Wilson whose work allegedly explained the "λ-anomalies". But it has been proven by Mnyukh and Vodyanoy [7] that all the theoretical efforts to account for the appearance and shape of the "heat capacity λ-anomaly" in liquid *He* are hopeless due to misinterpretation of its physical nature: it represents the *latent heat* of the liquid – liquid (non-"critical") phase transition, and not its *heat capacity* (more about it will be said in Section 6).

Which other phenomena, beside the *liquid – gas* "critical point" phase transition, are usually regarded "critical phenomena" in the scientific literature and among the scientific community at large? In case of solid – solid phase transitions their temperature is usually called "critical temperature" or "critical point" and marked $T_C$ whether it involves crystal structure, ferromagnetism (where it is also called "Curie point"), ferroelectricity, superconductivity, or anything else. Does that mean all those phase transitions are "critical phenomena"? In that regard a few relevant books were examined.

● **Statistical Physics, by L D Landau and E M Lifshitz** [1]. Sorting out all phase transitions into *first order* and *second order* by Landau was described. The *second order* transitions are certainly belong to "critical phenomena". The terms "second order" and "critical phenomenon" are probably equivalents, but such a statement in the relevant literature seems to be absent. Their equivalency is an important matter, since *second-order phase transitions do not really exist.* The chronological steps leading to that conclusion are given in [7]. That conclusion was made in spite of the fact that from 1934 the existence of such transitions became an unbreakable dogma. It was a result of comprehensive studies of solid-state phase transitions, initiated in 1960's, which have lead to finding their universal

mechanism described in Section 4.

As for the first-order transitions, Landau did not express any interest in their unknown molecular mechanism, just calling them "usual" and listing some of their properties. One of those properties was "overheating or undercooling is possible". In other words, they do not have a critical point and, therefore, are not a critical phenomenon. The listed properties of first-order phase transitions and those of the nucleation-and-growth (Section 4) are the same. The two concepts are identical.

● **Phase Transitions, by R Brout** [8]. *All* phase transitions are cooperative critical phenomena treatable by statistical mechanics. The theoretical goal was to justify the fact of phase transitions and account for their "singularities". Brout rationalized that at critical points the "infinite fluctuations give rise to singularity in full energy which, in turn, give rise to the heat capacity λ-anomaly". Theoretical models of the phase transitions, mostly Ising model, were used. In case of ferromagnetism the Ising model consists of a crystal lattice with its sites occupied by spins of two orientations, either 'up' or 'down', flipping between them arbitrarily. Due to interaction of the neighboring spins, small patches of parallel spins arise and grow as the temperature is decreasing toward the critical point $T_C$, at which the whole crystal somehow becomes magnetized[*].

When the real molecular mechanism of a ferromagnetic phase transitions [9,10] is invoked, it becomes evident that ferromagnetism cannot be approximated by Ising, Heisenberg, or any other theoretical model. The real mechanism is not based on fluctuations and it is not *cooperative* by its nature (Section 7). The theory is useless.

● **Introduction to Phase Transitions and Critical Phenomena, by R. Stanley** [5]. The title of the book leaves an impression that phase transitions in general are "critical phenomena". The content of the book in many respects is similar to the book just considered. All *solid-solid* phase transitions are deemed to be a *cooperative,* treatable by the methods of statistical mechanics. The Landau theory is discussed, but it is not specified that the theory is only about second-order phase transitions. The "first order" phase transitions are not mentioned. Only two phase transitions, *liquid – gas* and *ferromagnetic,* are chosen to be analyzed. They are claimed to be similar and typical for other phase transitions, such as in ferroelectrics and superconductors.

The central premise of the similarity between the *liquid – gas* and *ferromagnetic* phase transitions is erroneous, considering that ferromagnetic phase transitions are represented by the Ising model which does not approximate reality. As already pointed out, the model represents a *cooperative* process, while a real ferromagnetic transition is not. A real ferromagnetic phase transition proceeds by *local molecular rearrangement* at the interface between coexisting phases (Section 7).

● **Modern Theory of Critical Phenomena, by S K Ma** [11]. The claim to be modern was based on the K. Wilson's

---

[*] While the patches are magnetized, the crystal as a whole would not.

theoretical contribution – renormalization group approach to scaling. The Wilson theory was limited by second-order phase transitions, but that fact was quickly forgotten by everyone. The critical phenomena mentioned in the book are the phase transitions: *liquid-gas, ferromagnetic, antiferromagnetic, ferroelectric, in liquid He, in superconductors, in liquid and solid binary systems.*

Again, the ferromagnetic case was regarded representative of all of them. The introductory chapter describes a ferromagnet from $0 \, K$ up to its phase transition to the paramagnetic phase at critical point $T_C$. It goes as follows. At $0 \, K$ all spins have the same direction, so the ferromagnet is magnetized to saturation. With the temperature rising, some spins become randomly oriented. Their quantity increases with temperature until all of them are disoriented at $T_C$ and above it. Therefore, nothing happens at $T_C$ itself, which is why the *ferromagnetic – paramagnetic* phase transition is called *second order*.

That presumed description, tailored to justify using statistical mechanics methods, reveals the main problem of the theory: it has nothing to do with reality. It suffices to note that a ferromagnetic crystal has domain structure at all temperatures from $0 \, K$ to $T_C$, every domain being a single crystal magnetized to saturation. There are no randomly oriented spins between $0 \, K$ and $T_C$. Also, there is no reason for spins of all domains to be oriented in the same way at $0 \, K$. And there is no reason for all spins in a domain being parallel at $0 \, K$: why, for example, not to be in a helical disposition?

# 4. The Universal Mechanism of Solid-Solid Phase Transitions

A time ago the common belief was that some phase transitions are "structural", meaning representing a change of their crystal structure, and other (ferromagnetic, ferroelectric, order-disorder, normal-superconducting) are not. Those "non-structural" were declared "second order". Then some strange "magnetostructural" phase transitions were experimentally found. Structural changes "accompanying" other non-structural transitions were recorded as well. We contrast that with the statement that *all* phase transitions in solids are *structural* and have the same *nucleation-and-growth molecular mechanism*. In fact, knowledge of that structural mechanism is equivalent to a basic understanding of the nature of all phase transitions and other reactions in solid state. A frequent misconception that phase transitions are *accompanied* by changing of the crystal structure must be reversed. It is change of crystal structure which *is accompanied by* the changing of its properties, whether it is ferromagnetic, ferroelectric, electric conductivity, optical, etc.

The molecular mechanism of solid-solid phase transitions was disclosed in late 1960's by the present author after the discovery of the phenomenon of crystal growth in a single crystal medium (Fig 2). It was presented in detail in [12] and its universal character was then confirmed in ensuing

publications. Considering its key role in this article and that it is still far from general acceptance, its abbreviated description is given below. It is named *nucleation-and-growth*.

**Figure 2.** An example of phase transition in a transparent single crystal of *p*-dichlorobenzene. It is a growth of well-shaped single crystal of the new (higher-temperature) phase within the original lsingle-crystal phase. The transition started from a visible crystal defect at $T > T_0$ by several degrees. There is no rational crystallographic orientation relationship between the original and new phases

## 4.1. Nucleation in Microcavities

Nucleation of the alternative phase in solid-state phase transitions differs in all respects from the theoretically-born fluctuation-based statistical picture described by Landau and Lifshitz (Section 150 in [1]). In reality the nucleation sites are located in specific crystal defects – microcavities and microcracks of a certain optimum size. Every site contains information on the actual temperature of its activation. These temperatures are not the same in different defects, but nucleation in a given crystal is *predetermined* as for the location of the defect and its encoded temperature. It would not occur at all in perfect crystals, but such crystals are extremely rare.

## 4.2. Hysteresis

Every nucleus is activated only *after* the temperature $T_0$ marking the equality of free energies of the phases, $F1 = F2$, has been passed. The nucleation lags are inevitable in both directions of phase transitions. They are the only cause of *hysteresis*. Hysteresis is a necessary component of the phase transition mechanism. Its value is not exactly reproducible when transition was initiated by different nucleation sites. At least some finite hysteresis is inescapable. An "ideal phase transition" (*i. e.*, without hysteresis) cannot occur, since its thermodynamic driving force $\Delta F = F1 - F2 = 0$ at $T = T_0$.

Along with the hysteresis of initiation of the new phase there is also a smaller, but consequential, secondary hysteresis of sustaining the transition when the two phases already coexist.

### 4.3. Temperature of Phase Transition. Range of Transition

Temperature of phase transitions is regularly treated in scientific literature as $T_C$, calling it "critical temperature" (or "Curie temperature" or "critical point") – the term designated for "critical" phenomena. This is a consequential mistake, a source of countless misinterpretations of the nature of phase transitions. When the value of a physical property $P$ shows a gradual change "in the vicinity of transition" on the experimental $P(T)$ charts, it is interpreted as a pre-transition effect. The phase transition looks complicated. Location of the phase transition temperature becomes a subject of guessing, as a quotation in Section 1(c) illustrated, and the curves become a subject of mathematical delineation and interpretation.

The cause of these seeming complications is rooted in the fact that *nucleation-and-growth* phase transitions are always spread over a *range of transition* due to the hysteresis, wide or narrow, of two levels. The *hysteresis of initiation* usually involves nucleation from many sites of different encoded starting temperatures. Each emerging interface is subjected to the *hysteresis of sustaining* that inhibits its free propagation. Their combination results in spreading the transition over a temperature range, which is located *after* $T_0$ has been passed. In the *range of transition* the two phases coexist, while their ratio changes by local rearrangement at their interfaces.

### 4.4. Rearrangement at *Contact* Interface

A solid-state phase transition is intrinsically a *local* process. It proceeds by "molecule-by-molecule" structural rearrangement at interfaces only, while the bulk of both the original and emerged phases remains static (Fig.3). No "jumps" of macroscopic quantities or properties occur. The seeming "jumps" appearing on experimental recordings are simply the differences between the crystal structures or their physical properties when the transition range is either narrow, or passed quickly, or both. Coexistence of the phases during phase transition is self-evident. Simple observation of simultaneously present phases is a solid proof of the nucleation-and-growth phase transition.

### 4.5. Epitaxial Type of Nucleation-and-Growth Phase Transitions

Fig. 3 illustrates a case when the crystal orientations of the phases are not related. This is because the nuclei initiating the phase transition in the crystal defects are, in principle, oriented arbitrarily. However, a structural *orientation relationship* is frequently observed. These transitions, usually between similar structures, materialize by nucleation and growth as well, no matter how analogous the structures could be. The cause of the orientation relationship in these cases is *oriented nucleation* (epitaxy). They are *epitaxial* phase transitions.

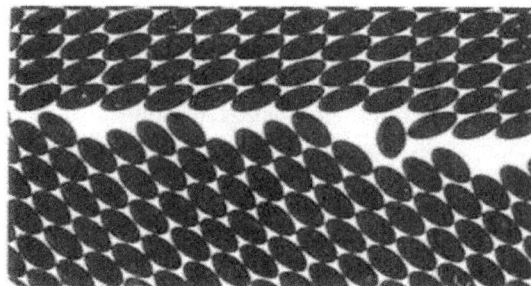

**Figure 3.** Model of *stepwise* (or edgewise) molecule-by-molecule rearrangement at the *contact* interface. The process is located at "kinks" on the crystal faces of the emerging phase (upper part) and proceeds by building it up, layer-by-layer, that way, while the original phase is dissolving. No structural orientation relationship between the phases is necessary. Every completed molecular layer requires nucleation to initiate the next layer

There are two circumstances when it takes place. One is when crystals have a pronounced layered structure. There the crystal layers in both phases are almost the same, the difference being in the layer stacking. Nucleation occurs in the tiny interlayer cracks. The nucleation is *epitaxial* due to the orienting effect of the "substrate" – the opposite surface of the crack. The layers after the transition retain their original orientation.

Another case of the *epitaxial* nucleation is when the unit cell parameters of the polymorphs are extremely close (roughly less than 1%) even in non-layered crystals. That condition is met, for example, in *paramagnetic–ferromagnetic* phase transition of iron (Section 7).

*Epitaxial* phase transitions exhibit themselves in a specific way: the original crystal orientation is preserved, the Laue X-ray patterns of the phases appear almost identical, the temperature hysteresis can be small and easily missed, so is the latent heat, etc. It is the *epitaxial* phase transitions that are taken for "second-order" or "weakly first-order" and improperly treated as "critical phenomena".

### 4.6. On "Weakly First-Order" Phase Transitions

The *nucleation-and-growth* molecular mechanism is the only alternative to the "critical phenomena". Its macroscopic indicators are: *phase coexistence, interfaces, range of transition, hysteresis*. Observation of any of them is sufficient to identify it, disqualifying the phase transition as being a "critical phenomenon". This mechanism is the equivalent of the *first-order* transitions in the Landau first- and second-order classification. However, second-order transitions, while they would be a true critical phenomenon, do not exist [7], which is partially recognized by Ginsburg *et al.* [13] with the admission that they are extremely rare. In the effort to save the Landau theory and the theory of critical phenomena, the additional category – "weakly first order" phase transitions – was introduced with an arbitrary claim that they can be treated as second-order. Their features – small hysteresis, short range of transition, structural similarity – are those of the *epitaxial* nucleation-and-growth.

The fundamental error of that approximation is that no matter how much "weakly" the first-order transitions are, *they are not a cooperative process*. They are still nucleation and growth by rearrangement at interfaces. Treating them by methods of statistical mechanics is unconditionally erroneous.

## 5. Lambda-Transitions

### 5.1. Puzzle is Solved, but Not by Theory

The sharp peaks reminiscent to letter $\lambda$, recorded in heat capacity measurements over the temperature ranges of solid-state phase transitions challenged the theoretical physicists to explain their origin. The first $\lambda$-peak was observed by Simon in $NH_4Cl$ phase transition [14]. Later on, this was repeated by many different authors and numerous other cases were also reported. Thus, more than thirty experimental $\lambda$-peaks presented as "Specific heat $C_P$ of [substance] *vs.* temperature $T$" were reproduced in [15]. P.W. Anderson wrote [16]: "*Landau, just before his death, nominated* [lambda-anomalies] *as the most important, yet unsolved problem in theoretical physics, and many of us agreed with him... Experimental observations of singular behavior at critical points* [were] *multiplied as years went on... For instance, it have been observed that magnetization of ferromagnets and antiferromagnets appeared to vanish roughly as* $(T_C\text{-}T)^{1/3}$ *near the Curie point, and that the $\lambda$-point had a roughly logarithmitic specific heat* $(T\text{-}T_C)^0$ *nominally*". Feynman stated [17] that "*One of the challenges of theoretical physics today is to find an exact theoretical description of the character of the specific heat near the Curie transition - an intriguing problem which has not yet been solved.*"

For now this intriguing problem has been solved [12,18,7], but not by a theory based on statistical mechanics. The solution was already contained in the experimental results of the 1942 publication by Dinichert [19], but overlooked since.

### 5.2. Classical Case of $NH_4Cl$

The canonical case of "specific heat $\lambda$-anomaly" in $NH_4Cl$ around -30.6°C is of special significance. It was studied for years. It was the only example used by Landau in his original articles on second-order phase transitions. In every calorimetric work a sharp $\lambda$-peak was recorded  (Fig. 4(b)). Neither author expressed doubts in a *heat capacity* nature of the peak. The case has been designated as a *cooperative order-disorder transition of the lambda type*. However, no one maintained that the $\lambda$-anomaly was understood.

Many of the calorimetric studies were undertaken well after 1942 when the experimental work on $NH_4Cl$ by Dinichert [19] was published. The transition was not a cooperative nature: microphotographs revealed coexisting phases separated by interfaces. The transition was spread over a temperature range where only mass fractions $m_L$ and $m_H$ of the two distinct L (low-temperature) and H (high-temperature) phases were changing, producing

"sigmoid"-shaped curves (Fig. 4(a)). The direct and reverse runs formed a hysteresis loop. The fact that the phase transition is first order was incontrovertible, but not identified as such.

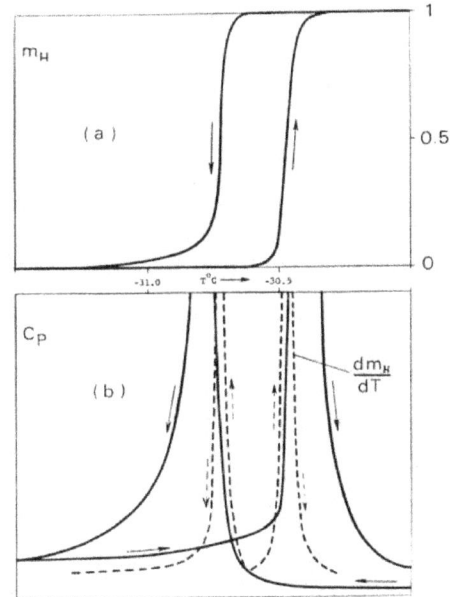

**Figure 4.**    Comparison of two plots representing the measurements through the range of $NH_4Cl$ phase transition, Plot 'a' by Dinichert represents the mass fraction $m_H$ in the two-phase L+H range of transition; $m_L + m_H = 1$. The plot 'b' (solid line [20]) is calorimetric $\lambda$-peaks. The two plots are shown in the same temperature scale to make it evident that shape and location of the peaks are very close to the first derivative (dashed curves) of $m_H$. The nature of the $\lambda$-peaks and their hysteresis becomes clear, since the absorption/release of the *nucleation-and-growth* latent heat is proportional to $dm_H / dT$

In Fig. 4 the Dinichert's data are compared with the calorimetric measurements (by other authors) showing "anomalies of heat capacity" (as they called the $\lambda$-peaks) and the hysteresis of the $\lambda$-peaks. Because of the hysteresis, it had to be evident at that point (but was not) that the $\lambda$-peaks cannot be a heat capacity, considering that heat capacity is a *unique function* of temperature. The graphs 'a' and 'b' are positioned under one another in the same temperature scale to reveal that the shape and location of the peaks are very close to *first derivative* of the $m_H(T)$ (dashed curves). It remains to note that *latent heat* of the phase transition must be proportional to $dm_H/dT$. Thus, the latent heat of the first-order phase transition is found, replacing the specific heat singularity. The mystery is solved.

### 5.3. "$\lambda$-Anomaly" is Latent Heat of Nucleation-and-Growth

Another way to prove the *latent heat* nature of the $\lambda$-peak is to use *differential scanning calorimetry* (DSC) to record it. That technique displays endothermic and exothermic peaks of latent heat with *opposite* signs in the chart recordings, If the $\lambda$-peak in $NH_4Cl$ phase transition is a *latent heat*, as was concluded above, the peak in a descending run must be exothermic and look downward. Our strip-chart recordings made with a Perkin-Elmer DSC-1B instrument immediately revealed that the peak acquires the opposite sign in the reverse run (Fig. 5). Its hysteresis was also unveiled.

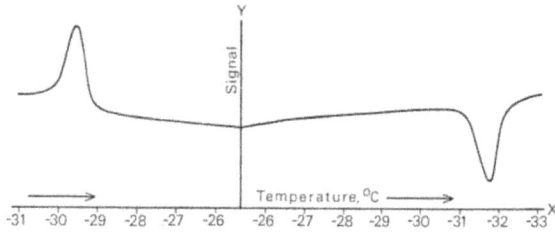

**Figure 5.**    The actual DSC recording of $NH_4Cl$ phase transition cycle [12], displaying temperature-ascending and descending peaks as endothermic and exothermic accordingly, thus delivering independent proof of a latent heat nature of the λ-peak. Hysteresis is seen too

## 6. λ-Anomaly in Liquid Helium

In Section 3 an authoritative statement was quoted about the great significance of the liquid helium "λ-transition" for the study of critical phenomena. Unfortunately for the theory of critical phenomena, the *"heat capacity anomaly"* used by Ehrenfest to introduce a *second-order* type of phase transitions does not exist. The subsequent theoretical works of many years on its shape delineation and analytic description does not have any scientific value. The consequential mistake was that the "λ-anomaly" in question (Fig. 1) is a *latent heat* of structural phase transition, and not a heat capacity.

It is not realistic to continue assigning a *heat capacity* to the λ-peak in liquid *He*, while the same peculiar peak in $NH_4Cl$ and many other cases is now proven to be a *latent heat*. In liquid *He* we deal with the *liquid polymorphism*. At any given moment liquid consists of tiny clusters of approximately closely packed particles. A certain kind of a *short-range* order within the clusters is preserved. Change in the manner the particles are packed in the clusters with temperature (or pressure) is similar to polymorphic phase transitions in crystals.

In the measurements of heat effect in the liquid *He* phase transition an one-way adiabatic calorimetry was used. There is no doubt that an appropriate calorimeter (such as DSC) will produce the peak in cooling and heating as looking in opposite directions (as in Fig. 5), thus confirming its *latent heat* nature. Simultaneously, its hysteresis will be revealed too.

Surprizingly, the liquid *He* phase transition is not a critical phenomenon.

## 7. Ferromagnetic Phase Transitions

Magnetization of a ferromagnet and a ferromagnetic phase transition served to the theory of critical phenomena as the exemplary object to explore by the methods of statistical mechanics. However, the Ising and other models utilized by the theories cannot even remotely approximate the real molecular mechanism of magnetization. The real mechanism is not a cooperative statistical process of flipping spins from one direction to the other. And it is not based on spin interactions.

A detailed new explanation of spontaneous magnetization, the domain structure and the magnetization process has been presented by this author in the books [12,21] and articles [9,10]. A very realistic assumption that *spin direction is fixed in the structure of its atomic carrier* easily solves all problems of the current theory of ferromagnetism, eliminating its main attributes – Heisenberg theory, anisotropy energy and Bloch wall. It accounts for all its manifestations: its phase transitions, molecular mechanism, the diversity of magnetic structures, hysteresis loops, and all the rest.

Magnetization and ferromagnetic phase transitions do not occur by flipping or rotation of spins in the same structure. The source of spontaneous magnetization is *crystal field* that sets up the orientations of crystal particles and, therefore, the orientation of their spins. Thus, ferromagnetic phase transition and magnetization (spin reorientation) can materialize only by structural rearrangements. It proceeds by propagation of *contact* interfaces, where magnetic particles change their orientation by relocation, one particle at a time, in accordance with the universal *stepwise* mechanism of solid-state phase transitions outlined in Section 4.4.

Ferromagnetic phase transitions by moving interfaces have been observed and photographed by a number of different authors. This fact alone already reveals that the phenomenon is not of a cooperative nature and, therefore, not "critical". The work by Novoselov *et al.* [22] is to be specifically mentioned. The authors were able to record a domain interface with the resolution at the molecular level. The domain interface propagated "via clear quantized jumps matching the lattice periodicity". Also observed were "kinks" running along the interface. In other words, it was a *stepwise rearrangement at the contact interface* shown in (Fig. 3).

The last point to clarify is about the accompanying "singularities". Late R.P. Feynman, a Nobel Prize Laureate, raised that problem [17]: *"One of the challenges of theoretical physics today is to find an exact theoretical description of the character of the specific heat near the Curie transition - an intriguing problem which has not yet been solved. Naturally, this problem is very closely related to the shape of the magnetization curve in the same region".* The problem has been solved, but not by the theory of critical phenomena. The heat effect in question is the *latent heat* of the *structural* magnetic phase transition. It is incontrovertible: ferromagnetism and ferromagnetic phase transitions are not a critical phenomenon.

## 8. Ferroelectric Phase Transitions

The phenomenon of ferroelectricity is very much analogous to ferromagnetism. Electric dipoles in ferroelectrics have the same role as spins in ferromagnets do. Nevertheless, the only recognized attempt of a theoretical account for ferroelectricity, represented by the voluminous book by Lines and Glass [23], did not reflect that analogy. While ferromagnetic phase transitions are believed to occur by spin flipping in the same crystal lattice, ferroelectric phase transitions are claimed to occur by a sudden change of the crystal structure. That sudden change supposedly materialized by a displacement of electrically charged particles in the original crystal lattice. That happens at a critical temperature $T_C$ when the amplitude of one of the crystal "soft" optical modes becomes "frozen". The authors related the ferroelectric *soft-mode* phase transitions "to the more general field of structural transitions and even to critical phenomena in general".

There are, however, fundamental inconsistencies regarding that "critical phenomena" proclamation. (1) Such an instant jump does not fit the definitions of a critical point and, even worse, the critical point does not exist at all due to the hysteresis observed experimentally. (2) The ferroelectric *soft-mode* phase transitions coexisted in that theory with the recognition that "most ferroelectric phase transitions are not of second order but first". (3) The *soft-mode* concept *per se* was found invalid and disappeared from scientific literature, except when it is related to ferroelectric phase transitions.

The close parallelism of ferromagnetism and ferroelectricity is not accidental: both are the effect of the same general cause. *The disposition and orientation of spins and electric dipoles are arranged by the crystal forces which combine their carriers into a crystal.* This approach has been presented initially in 2001 [12], and then, in more complete form, in [9]. The origin of ferroelectrics, their classification, formation of their domain structure, their stability, difference from pyroelectrics, formation of their hysteresis loops, parallelism with ferromagnetism, and more, were presented coherently in terms of the universal *nucleation-and-growth* mechanism (Ferroelectric state and phase transitions, *in* [21], p. 117-132). Ferroelectricity and ferroelectric phase transitions are not a critical phenomenon.

# 9. Order – Disorder Phase Transitions

The term *order – disorder* denotes a kind of phase transitions when particles of the higher-temperature phase become rotating, while being arranged in a positional 3-D crystal order. The transition in $NH_4Cl$ is one of them. For a while the common opinion was that such rotation starts in the same crystal lattice. However, phase transitions *order→disorder* do not occur by rotation of particles in the original crystal. The experimental results by Dinichert show transition in $NH_4Cl$ proceeding by a gradual increasing of the *quantity* of one of the *coexisting* crystal phases at the expense of the other. The work also reveals *hysteresis* of the transition. All that now points to the *nucleation-and-growth* molecular mechanism. Phase transition in $CBr_4$ is another example. Observation with an optical microscope showed growing single crystals of the "disordered" phase in the medium of the original ("ordered") crystal phase [24]

(Fig. 6).

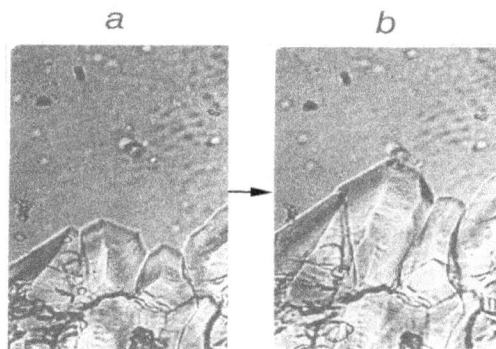

**Figure 6.** Order-disorder phase transition in $CBr_4$. (a,b) Conglomeration of single crystals of "disordered" phase with rotating molecules grows within the crystal medium of the original "ordered" phase. The natural facing is evident. The two phases coexist, while all phase rearrangement occurs at the interfaces. This is nucleation-and-growth, and not a cooperative critical phenomenon

Changes of the lattice parameters in many order-disorder phase transitions are small (as, for example, in the ferromagnetic phase transition of iron), resulting in erroneously labeling them *second order*, which means being a critical phenomenon. That led to the following footnote in the Landau & Lifshitz book [1]: *"There are claims in the literature about relation between second-order phase transitions and appearance of rotating molecules in the crystal. Such view is erroneous ... ".* The real cause of that wrong classification was that those phase transitions materialize by the *epitaxial* kind of the nucleation-and-growth. *Order – disorder* phase transitions are not a critical phenomenon either.

# 10. Normal – Superconducting Phase Transitions

Immediately after the discovery of superconductivity in 1911 by Kamerlingh-Onnes the question of how superconducting state emerges from higher-temperature "normal" non-superconducting state arose. It had to be first established whether a superconducting phase has its individual crystal structure. Onnes was inclined to view the transition as polymorphic change. That possibility, however, was rejected in1924 by an X-ray experiment: no change in the diffraction patterns of lead was detected. The conclusion about crystallographic identity of pre-superconducting and superconducting phases was an error caused, probably, by the imperfection of the technique available at that time. It served as a basis to erroneously categorize these transitions as *second order* – the belief lasting up to the present time.

All superconductors known at that time are now recognized by most to exhibit *first order* phase transition, basically because they do not fit to be second order. The 1971-1972 articles by Ott [25-27] are to be especially mentioned, where it was reported that single crystals of lead, aluminum, zinc and gallium undergo dimensional and volume changes upon their *normal – superconducting* phase transitions. This alone is incontrovertible proof that they are not a critical phenomenon.

Examination of the experimental literature on "high-temperature" superconductors [7] revealed that certain structural changes resulting from *normal–superconducting* phase transitions do occur, even though sensitive techniques were needed to catch them. Ample evidence of that was presented in the book "Lattice Effects in High-$T_C$ Superconductors" [28]. These changes are reported as lattice distortions, thermal expansion jumps, atomic displacements from regular lattice sites, local displacements from average crystal structure, changes of some inter-atomic distances, etc. In the following years the reports about a "structure anomaly", "lattice distortion", or alike, at $T_C$ in different high-temperature superconductors continued to appear [29-32], suggesting that the phenomenon is general to all superconductors.

Structural changes in superconductors and non-superconductors alike do not occur by displacements, distortions, deformations or jumps. Notwithstanding how minute the structural changes could be, they are still macroscopic and, therefore, not allowed by thermodynamics. Any phase transition is a replacement, and not modification, of the old structure. The normal – superconducting phase transitions materialize by nucleation-and-growth. All accounts indicate that they are of the epitaxial type. They are not a critical phenomenon.

# 11. Non-Critical Opalescence

### 11.1. Misplacement of Transition Temperature Makes Opalescence Look "Critical"

The arbitrary placement of the "critical point" $T_C$ at the top of a λ-peak turns phase transitions into extremely complicated, even mysterious phenomenon. They are perceived as starting well *before* the $T_C$ and completing well *after* it. The pre-transition and post-transition processes in a crystal medium became a central problem of critical phenomena theory.

Realization that a phase transition proceeds over a temperature range *after* the equilibrium temperature $T_0$ has been passed (Section 4.2) eliminates the mystery. The universal *nucleation-and-growth* mechanism brings about a simple explanation. The peaks of light scattering form *after* $T_0$ has been passed. A sharp increase in the light scattering is caused by the mass nucleation of the new phase. Hysteresis of the peaks of light scattering has indeed been observed in the experiments.

### 11.2. Static Source of the Opalescence

The problem of narrow peaks of light and neutron scattering that centered, as was believed, at the "critical temperature $T_C$" was called *central peak problem*. Regarding light scattering alone, extensive literature was devoted to this

phenomenon called *critical opalescence*. A notorious "λ-anomaly" can be recognized in the central peak (Fig. 7). The significance of this phenomenon for theory of critical phenomena was expressed in the 1979 article "Quasielastic Light Scattering Near Phase Transitions" by Lyons and Fleury [33]: "*Our understanding of structural phase transitions has evolved in the last decade through several stages from the simple soft mode and mean field theories to the modern coupled-mode, renormalization group and dynamic scaling ideas. Crucial to this evolution has been an increasingly detailed interpretation of the experimental data, particularly those resulting from scattering experiments*". A disconcerting circumstance, however, was that by 1979 it was already a well-proven fact that the central peaks were caused by scattering from *static* centers. At that stage Lyons and Fleury were compelled to suggest that the scattering centers were "long-living clusters" rather than entropy fluctuations. But failure of the dynamic theories was not recognized.

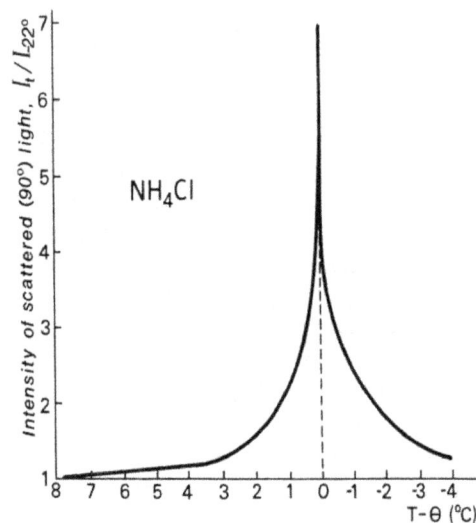

**Figure 7.** The apparent "critical opalescence" peak of light scattered at 90° to the primary light beam by a *NH₄Cl* single crystal during its low-temperature ($T_0$ = -30.6°C) phase transition (Shustin [34]). Presentation of the temperature scale as a distance from the position of the peak maximum obscures that the peak is not located at $T_0$

The 1981 review by the same authors [35] revealed a certain evolution in representation of the subject: though the *critical phenomena* approach was preserved, a separate section entitled "Static and dynamic central peaks" appeared. There it was stated that the first instance of the central peak of light scattering reported for quartz by Yakovlev *et al*. [36] was later proven [37] to be due to "scattering from static microdomains, rather than to true critical opalescence", and that "in most other cases confusion still persists...". Also mentioned was $K_2PO_4$ where the central peak has been observed [38], but shown [39] to be "largely of static origin". Omitted from this account was the case of *NH₄Cl* (Fig. 7) usually considered a valuable asset in illustrating critical phenomena: the scattering centers there also were static [40].

Other believers in critical phenomena encountered the

same problem. Ginzburg et al. in their review on light scattering [41] listed more examples of the central peaks caused by static scattering centers: $SrTiO_3$, $Pb_5Ge_3O_{11}$, $KH_3(SeO_3)_2$. It was explained that "extraction of the part due to thermal fluctuations from the total scattering intensity is generally a rather hard task". But no evidence that such a part existed was presented.

### 11.3. Explanation was Ignored

The explanation of central peaks was proposed by Bartis as early as 1973 in the articles "Critical Opalescence in Ammonium Chloride" and "The Transitional Opalescence of Quartz" [40,42]. Some assumptions leading to this explanation were not quite valid. A phase transition was considered a three-step process in which only the intermediate stage was a temperature range of two-phase coexistence. But it was sufficient to advance the following idea: *"To understand the increased scattering of light we focus our attention on the interface between the two crystalline forms in the intermediate stage. Inasmuch as the two forms have substantially different properties, light incident on the interface is bound to experience some scattering."*

Here is a synopsis of Bartis' reasoning. Interest in critical phenomena had led to an odd turn of events due to "the discovery of a discontinuous intermediate stage in a score of transitions previously believed to be second order". The intermediate stage should produce a sharp rise in the scattered light that could easily be mistaken for critical opalescence. Ginzburg, and then Yakovlev et al., had made this mistake when they attributed the opalescence of quartz during its α – β transition to critical fluctuations. Even though the Ginzburg's theory predicted ~$10^4$ increase in the light scattering, as was actually observed, it would be proven later that the effect had nothing do with critical fluctuations. Two forms of quartz coexist over 1°. The transitional opalescence in quartz was caused by light scattering by the interfaces.

The Bartis' account, however, was inconsistent with the orthodox theory developed over previous years. The almost self-evident solution was ignored. It was neither discussed, nor taken into account even in the subsequent publications devoted to the central peaks in quarts and $NH_4Cl$. The Bartis' articles were not included in the comprehensive list of about 700 references accompanying the 1981 review by Fleury and Lyons on the topic [35]. The cause of the central peaks had already been named, but the search for it was continuing. Lyons and Fleury stated [33] that further theoretical work is needed. Ginzburg et al. [41] modified the theory to include "static inhomogeneities and defects" as the primary contributors to the central peaks. The new theory involved critical fluctuations and did not mentioned nuclei and interfaces.

### 11.4. Sours of Light Scattering Seen in Microscope

(1) 1975. Durvasula and Gammon [43] pointed out that light scattering investigations failed to measure temperature

carefully and paid little attention that the phase transitions exhibit first-order behavior. In their examination of the phase transition in $KH_2AsO_4$ the light scattering measurements were combined with dielectric measurements and visual monitoring of emergence of the second phase (ferroelectric domains). Both heating and cooling runs were carried out. A hysteresis loop was detected, which meant absence of the critical point.

It was found that sharp increase in light scattering starts at that same moment when the domains appear in the illuminated area of the specimen. That was a direct demonstration that the source of the light scattering responsible for the central peak was the interfaces emerging in the two-phase coexistence range. However, the leading theorists in the field would continue using vague language discussing the "static origin" of the central peaks and speculating about the nature of the "long-living clusters" or "inhomogeneities". The nature of the "inhomogeneities" as a two-phase state had already been established, but Ginzburg et al. would suggest years later [41] that they are crystal defects.

(2) 1977. The detailed optical microscopic study of the $NH_4Cl$ phase transition by Pique, Dolino and Vallade [44] unambiguously identified the cause of the central peak. Their results completely substantiated the interpretation by Bartis. They stressed the importance of heterophase region (which they also called "mixed state" and "coexistence state"). "During the coexistence state ... there is such an intense light scattering that...no light remains in the [primary] beam of the *He-Ne* laser".

(3) 1979. A meticulous investigation of the α – β phase transitions in quartz has been carried out by Dolino [45]. Light scattering measurements were combined with optical microscopic observations. In the vicinity of the transition (~573°C) small temperature gradient was applied along the investigated quartz single crystal. A two-phase band moving from the α-end of the crystal to its β-end and in the reverse direction was observed. Dolino arrived at the conclusion that the light scattering centers were nuclei of the α-phase in the β-phase on cooling, and β-phase in α-phase on heating. The central peak resulted from light scattering by this heterophase band.

(4) The phase transition in a single crystal of β-paradichlorobenzene shown in Fig. 8 presents an opportunity to perform an imaginary experiment on light scattering. When a phase transition has the morphology like in Fig. 2, the appearance of a single crystal of the new phase would, obviously, start light scattering: this is why we can see it. This picture of phase transition is usually obscured by nucleation from multiple sites. The phase transition shown in Fig. 8 goes too quickly to reveal naturally shaped crystals of the new phase. It proceeds by moving an interface, seen as an opaque band, from left to right over the crystal. The band is a cloud of tiny crystals of the new phases. Recording of light scattering from a small area of the crystal would reveal a good "central peak" when the opaque band passes that area.

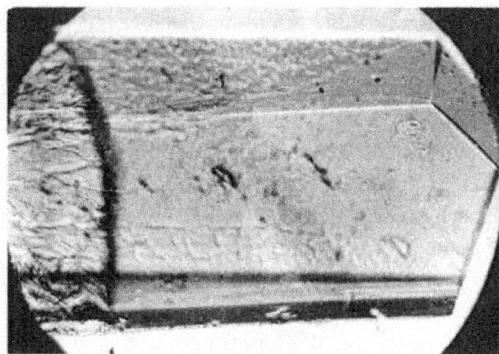

**Figure 8.** Single crystal of β-paradichlorobenzene in some intermediate stage of its phase transition. In this particular case the transition proceeds by moving interface, seeing as an opaque band, from left to right. If light scattering is measured from a small area of the crystal, the recording will produce a "central peak" when the interface passes that area

### 11.5. Theory *vs* Evidence

Light scattering "central peaks" in crystal phase transitions have underwent their own transition - from best friends of critical fluctuations to their worst enemies. The 1983 articles collection "Light Scattering Near Phase Transitions" [46] reveals it best. The leading article is characteristic of all the others in the book and sufficient to illustrate the point.

The leading article was the 155-page "General Theory of Light Scattering Near Phase Transitions in Ideal Crystals" by Ginzburg, Sobyanin and Levanyuk (GSL) [13]. The authors disregarded the works on phase transitions by nucleation and growth and works on critical opalescence described in Sections 11.3 and 11.4. The theory was to be applied to *second-order* phase transitions, but "[they] occur, strictly speaking, only in some exceptional cases". The predicament was resolved by a presumption that the Landau theory is applicable to "weakly" first-order transitions (Section 4.6). GSL decided that these transitions occur by "inner deformations" and nominated them to be the "order parameter". Fluctuations of that fictitious parameter were the central idea of their theory. Thus, the theory by GSL is applicable neither to most crystal phase transitions because they are "strongly" first-order, nor to the "weakly first-order" because they are not different from "strongly" ones (see Section 4.6), nor to the "pure" second-order transitions because they are nonexistent.

But there are more to be said about the GSL work. They did not claim any more that increasing in light scattering due to critical fluctuations must be as great as $10^4$ times. Now they maintained that the effect is extremely weak and has to be extracted from the "integral intensity" caused by crystal defects. They acknowledged that "no critical region has been yet observed with confidence" with the exception of the "critical points in magnetics". But this exception should be excluded (Section 7). GSL formulated the result of their work as follows: "The existing experimental data on the integral intensity of the thermal (molecular) light scattering near phase transition points in solids are neither reliable nor full enough to be compared with the above theoretical

predictions."

The objects, which the GSL theory was to be applied to, did not exist in reality. The light scattering by nuclei and interfaces in real phase transitions, described in previous sections, was not mentioned. The GSL work was dishonest and useless.

Such is the instructive story of the "critical opalescence" in solid-state phase transitions. The experimentally established source of the "central peaks" has been met with resistance by the prominent authors of the insolvent theories. Searching for solution of an already solved problem was continuing. Critical opalescence is still cited as a prominent feature of critical phenomena.

### 11.6. "λ-Anomaly" of Neutron Scattering

The central peaks of neutron scattering do not need to be discussed in detail. Their story is a carbon copy of what happened to the central peaks of light scattering.

- Original excitement over the discovery of a phenomenon that seemed to offer an intimate insight into the critical dynamics of solid-state phase transitions. The first observation of a neutron scattering central peak was reported in 1971 [47], 15 years after the first light scattering peak was reported. At that time the virtual "non-critical" cause of the light scattering peaks had already been established.
- Subsequent disappointment after finding that the neutron central peaks were a scattering from static centers.
- Search for the nature of the scattering static centers anywhere except where it has been experimentally observed in light scattering: nuclei and interfaces in the two-phase range.
- Dealing with the *first-order* phase transitions as if they were *second-order*.
- Desperate experimental search for a "dynamic" component in the "largely static" central peaks. The scope of experimentation was limited by the higher cost of the neutron scattering experiments, as well as by problems with having sufficiently intense neutron sources. As a result, the experimental data were less reliable than in light scattering, thus leaving a wider field for speculations.
- Statements that more work and time is needed to understand the origin of neutron central peaks.

Like in the light scattering, the neutron scattering was caused by nuclei and interfaces.

## 12. Conclusions

Over the 20th century and up to now the theory of critical phenomena has grown into a voluminous branch of theoretical physics. Most of that time the physical nature of its objects, which are basically phase transitions, was still unknown, so the theorists analyzed their models, such as Ising model of ferromagnetic phase transition. The models were of a cooperative kind, based on ideas of thermal fluctuations, allowing them to be treated by statistical mechanics.

When the molecular mechanism of solid-state phase transitions became ultimately known, it turned out to be in a striking difference with those models. It does not involve cooperative fluctuations. It is based on the nucleation-and-growth principle. It is general to all types of phase transitions: structural, ferromagnetic, ferroelectric, order – disorder, normal – superconducting, so none is a subject of statistical mechanics. The theoretical models cannot even remotely approximate them. The theoretical models and the real mechanism are antonyms.

As for the "λ-anomalies" and the "critical opalescence" (and a host of the effects caused by the opalescence), they have been incorrectly identified and have now received simple "non-critical" explanations. To that it should be added that (a) the famous *"heat capacity λ-anomaly"* in liquid helium phase transition is shown to be *latent heat* of the structural phase transition, and (b) the classical "critical point" of *liquid – gas* phase transition should be striped of its rank "critical" by the reasons given in Section 2.2.

It looks like the theory of critical phenomena lacks an object it might be applied to.

---

# REFERENCES

[1] L D Landau, E M Lifshitz: Statistical Physics, Addison-Wesley (1958).

[2] K P Belov: Magnetic Transitions, Boston Tech. Publications (1965).

[3] A I Kitaigorodskii: Organic Chemical Crystallography, Consultants Bureau, New York (1961).

[4] I Z Fisher: Critical state: *in* Physical Encyclopedic Dictionary (Rus.), Vol. 2, 545-547 (1962).

[5] H E Stanley: Introduction to Phase Transitions and Critical Phenomena, Clarendon Press, Oxford (1971).

[6] G Jaeger: The Ehrenfest classification of phase transitions: Introduction and evolution, Archive for History of Exact Sciences 53(1), 51-81 (1998).

[7] Y Mnyukh, V. J. Vodyanoy: Superconducting state and phase transitions, American Journal of Condensed Matter Physics 7(1): 17-32 (2017).

[8] R Brout: Phase Transitions, University of Brussels, New-York – Amsterdam (1965).

[9] Y Mnyukh: Ferromagnetic state and phase transitions. American Journal of Condensed Matter Physics 2(5), 109-115 (2012).

[10] Y Mnyukh: Magnetization of ferromagnets, American Journal of Condensed Matter Physics 4(4): 78-85 (2014).

[11] S K Ma: Modern Theory of Critical Phenomena, Benjamin/Cummings (1976).

[12] Y Mnyukh: Fundamentals of Solid-State Phase Transitions, Ferromagnetism and Ferroelectricity, Authorhouse (2001) [or 2nd (2010) Edition].

[13] V L Ginzburg, A A Sobyanin and A P Levanyuk: General theory of light scattering near phase transitions in ideal crystals, *in* Light Scattering Near Phase Transitions, ed. H Z Cummins and A P Levanyuk, North-Holland (1983).

[14] F Simon: Analysis of specific heat capacity in low temperatures, Annalen der Physik, 68, 241 (1922).

[15] N G Parsonage and L A K Staveley: Disorder in Crystals, Clarendon Press (1978).

[16] P W Anderson: The 1982 Nobel Prize in Physics, Science 218, no. 4574, 763 (1982).

[17] R Feynman, R Leighton, M Sands: The Feynman Lectures on Physics, vol. 2, Addison-Wesley, New York (1964).

[18] Y Mnyukh: On phase transitions that cannot materialize, American Journal of Condensed Matter Physics 4(1): 1-12 (2014), *or* arxiv.org/abs/1310.5009.

[19] P Dinichert: La transformation du $NH_4Cl$ observee par la diffraction des rayons X, Helvetica Physica Acta 15, 462-475 (1942).

[20] R Extermann, J Weigle: Anomalie de la chaleur specifique du chlorune d'ammonium. Helvetica Physica Acta 15, 455-461 (1942).

[21] Concluding Papers of Yuri Mnyukh on Phase Transitions, DirectScientific Press, Farmington, CT, USA (2018).

[22] K S Novoselov, A K Geim, S V Dubonos, E W Hill, I V Grigorieva: Sub-atomic movements of a domain wall in the Peierls potential: Nature 426, 812 (2003).

[23] M E Lines, A M Glass: Principles and Applications of Ferroelectrics and Related Materials, Clarendon Press (1977).

[24] Y Mnyukh, N I Musaev: Mechanism of polymorphic transition from crystalline to the rotational state, Soviet Physics – Doclady 13, 630 (1969).

[25] H Ott: Anisotropy in the length change of gallium single crystals at their superconducting transition, Solid State Communications 9, 2225-8 (1971).

[26] H Ott: The volume change at the superconducting transition of lead and aluminum, Journal of Low Temperature Physics 9, 331 (1972).

[27] H Ott: Anisotropic length changes at the superconducting transition of zinc, Physics Letters 38, 83-4 (1972).

[28] Y Bar-Yam, T Egami, J M Leonand, A R Bishop: Lattice Effects in High-$T_C$ Superconductors, World Scientific, Santa Fe, NM (1992).

[29] A E Boehmer, F Hardy, L Wang, T Wolf, P Schweiss, C Meingast: Superconductivity-induced re-entrance of the orthorhombic distortion in $Ba_{1-x}K_xFe_2As_2$, Nature Communications 6 (2015), doi:10.1038/ncomms8911.

[30] H Maeta, F Ono, C Kawabata, N L Saini: A search for extra lattice distortions associated with the superconductivity and the charge inhomogeneities in the high-$T_C$ $La_{1.85}Sr_{0.15}CuO_4$ system, Journal of Physics and Chemistry of Solids 65, 1445-8 (2004).

[31] H Oyanagi, C Zhang: Local lattice distortion in superconducting cuprates studied by XAS, Journal of Physics: Conference Series 428, 12042 (2013).

[32] C J Zhang, H Oyanagi, Z H Sun, Y Kamihara, H Hosono: Low-temperature lattice structure anomaly in the LaFeAsO$_{0.93}$F$_{0.07}$ superconductor by x-ray absorption spectroscopy: Evidence for a strong electron-phonon interaction, Physical Review B 78 (2008).

[33] K B Lyons, P A Fleury: Quasielastic light scattering near phase transitions, *in* Light Scattering in Solids, ed. J. Birman and H. Cummins, Plenum Press, p.357 (1979).

[34] O A Shustin: Scattering of light in the phase transition of the NH$_4$Cl, JETP Letters 3, 320 (1966).

[35] P A Fleury, K B Lyons: Central peaks near structural phase transitions, *in* Structural Phase Transitions, ed. K. Muller and H. Thomas, Springer-Verlag (1981).

[36] I A Iakovlev, T S Velichkina, L F Mikheeva: Opalescence in the phase transition of quartz, Soviet Physics-Doklady 1, 215 (1956).

[37] S M Shapiro, H Z Cummins: Critical opalescence in quartz, Physical Review Letters 21, 1578 (1968).

[38] N Lagakos, H Z Cummins: Critical opalescence in quartz, Physical Review Letters 34, 883 (1975).

[39] L N Durvasula R W Gammon: Nature of the central-peak light scattering in potassium dihydrogen phosphate, Physical Review Letters 38, 1081 (1977).

[40] F J Bartis: Critical opalescence in ammonium chloride, Physics Letters 43a, 61 (1973).

[41] V L Ginzburg, A P Levanyuk, A A Sobianin, A S Sigov: *in* Light Scattering in Solids, ed. J.L. Birman and H.Z. Cummins, p.331, Plenum Press (1979).

[42] F J Bartis: The transitional opalescence in quartz, Journal of Physics C L295 (1973).

[43] L N Durvasula, R W Gammon: *in* Light Scattering in Solids, ed. M. Balkanski, R C C Leite and S P S Porto, p.775, Flammarion Press (1975).

[44] J P Pique, G Dolino, M Vallade: Optical microscopic study of the NH$_4$Cl phase transition with observations of slip bands, heterophase and domain structure, Journal de Physique 38, 1527 (1977).

[45] G Dolino, Journal of Physics and Chemistry of Solids 40, 121 (1979).

[46] Light Scattering Near Phase Transitions, ed. H Z Cummings, A P Levanyuk, North-Holland (1983).

[47] T Riste, E J Samuelson, K Otnes and J Feder, Solid State Communications 9, 1455 (1971).

www.ingramcontent.com/pod-product-compliance
Lightning Source LLC
Chambersburg PA
CBHW061326190326
41458CB00011B/3905